SIMPLY
QUANTUM PHYSICS

DK LONDON
Project Editor Miezan van Zyl
US Editor Karyn Gerhard
Art Editor Mik Gates
Designer Clare Joyce
Managing Editor Angeles Gavira
Managing Art Editor Michael Duffy
Production Editor Gillian Reid
Senior Production Controller Meskerem Berhane
Jacket Design Development Manager Sophia M.T.T.
Jacket Designer Akiko Kato
Associate Publishing Director Liz Wheeler
Art Director Karen Self
Publishing Director Jonathan Metcalf

First American Edition, 2021
Published in the United States by DK Publishing,
a division of Penguin Random House LLC
1745 Broadway, 20th Floor, New York, NY 10019

A catalog record for this book
is available from the Library of Congress.
ISBN 978-0-7440-2848-5

DK books are available at special discounts when
purchased in bulk for sales promotions, premiums,
fund-raising, or educational use.
For details, contact: DK Publishing Special Markets,
1745 Broadway, 20th Floor, New York, NY 10019
SpecialSales@dk.com

Printed and bound in China

www.dk.com

CONTENTS

THE QUANTUM WORLD

PRE-QUANTUM PUZZLES

THE WAVE FUNCTION

INTERPRETATIONS OF QUANTUM PHYSICS

QUANTUM PHENOMENA

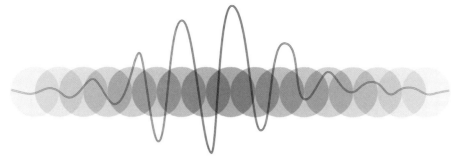

QUANTUM GRAVITY

QUANTUM BIOLOGY

CONSULTANT EDITOR
Dr. Ben Still is a prizewinning science communicator, particle physicist, and author. He teaches high school physics and is also a visiting research fellow at Queen Mary University of London. He is the author of a growing collection of popular science books and travels the world teaching particle physics using LEGO®.

CONTRIBUTORS
Hilary Lamb is an award-winning journalist and author, covering science and technology. She has written for previous DK titles, including *The Visual Encyclopedia, How Technology Works*, and *The Physics Book*.

Giles Sparrow is a popular-science author specializing in physics and astronomy. He has written and contributed to bestselling DK titles, including *The Physics Book, Spaceflight, Universe*, and *Science*.

THE

QUAN

WORL

T U M

D

Quantum physics describes the way the universe behaves on the very smallest scales. Far below the limits of even the most powerful microscopes, it governs the behaviors and interactions of atoms and the particles from which they are made—the fundamental building blocks of matter. Scientists only confirmed the existence of subatomic particles with J.J. Thomson's discovery of the electron in 1897, but the possibility that these tiny particles can sometimes behave like waves, which is key to the strange behavior of the quantum world, was only suggested by Louis Victor de Broglie in 1924.

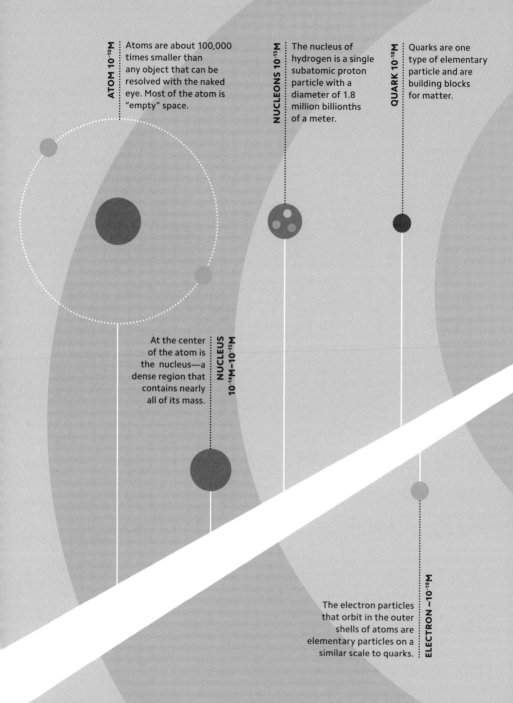

ATOM 10⁻¹⁰M — Atoms are about 100,000 times smaller than any object that can be resolved with the naked eye. Most of the atom is "empty" space.

NUCLEONS 10⁻¹⁵M — The nucleus of hydrogen is a single subatomic proton particle with a diameter of 1.8 million billionths of a meter.

QUARK 10⁻¹⁸M — Quarks are one type of elementary particle and are building blocks for matter.

NUCLEUS 10⁻¹⁴M–10⁻¹⁵M — At the center of the atom is the nucleus—a dense region that contains nearly all of its mass.

ELECTRON ~10⁻¹⁸M — The electron particles that orbit in the outer shells of atoms are elementary particles on a similar scale to quarks.

Quantum physics studies phenomena that occur at extreme measurements. Subatomic particles cannot be observed directly but they can be studied through experiments that observe their effects.

PLANCK LENGTH
10^{-35}M

This is the smallest unit of length possible in current physics theories. At lengths at or below the Planck length, current theories of physics break down and can no longer make sensible predictions.

VANISHINGLY SMALL

While the largest atoms have a diameter of about half a nanometer (billionth of a meter)—less than $1/100,000$th the width of a human hair—most of their volume is a sparse cloud filled with electrons around a dense central nucleus. Diameters of atomic nuclei are typically a few femtometers (million billionths of a meter), and it is usually at around these scales (and even smaller ones) that strange quantum behaviors become apparent. The smallest distance that makes physical sense is a Planck unit of length (see pp.140–41).

SOLID SPHERE MODEL

In 1803, John Dalton presented his theory that all matter is made from atoms—indivisible spheres that cannot be created or destroyed. However, atoms can be bonded or broken apart from other atoms to form new substances.

PLUM PUDDING MODEL

In J.J. Thomson's model, negatively charged electrons are dotted randomly throughout a sphere, which has a positive charge.

RUTHERFORD MODEL

Experimental evidence led Ernest Rutherford in 1911 to propose that the entire positive electric charge in an atom lay in a small, dense core and the electrons were imagined to orbit around this nucleus, like moons around a planet.

BOHR MODEL

To explain light absorption and emission by atoms, Niels Bohr developed a model in which electrons could orbit only in particular energy "shells."

THREE TINY PIECES

Atoms are the fundamental building blocks of large-scale matter—particles that were first thought indivisible and whose collective chemical and physical properties make them representative of one or another specific element. On a deeper level, however, all atoms are made up of a combination of three subatomic particles: positively charged protons and uncharged neutrons in a central nucleus, and negatively charged electrons orbiting in more distant clouds (see p.31), which allow atoms to bond with other atoms.

Electron clouds

In modern models of the atom, electrons are not solid spheres orbiting a nucleus at a fixed distance. Instead, they are represented as clouds in which electrons are most likely to be found if looked for.

There is an electromagnetic attraction between negatively charged electrons and positively charged protons in the nucleus.

ATTRACTION

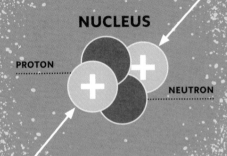

QUANTUM MODEL

NUCLEUS

PROTON

NEUTRON

In the atom, the number of negatively charged electrons balances out the number of positive protons in the nucleus.

ELECTRONS

Clouds of various shapes represent the orbitals in which electrons are most likely to be found.

ELECTRON CLOUD

PARTICLE ZOO

While electrons are truly elementary
particles, which cannot be divided
any further (and part of a family of
particles called the leptons), protons
and neutrons are made up of three
even smaller particles called quarks
(see p.122). Particles formed by
groups of quarks are collectively
known as hadrons, which are
subdivided into baryons (made
up of triplets of quarks) and
mesons (made up of a paired
quark and antiquark particle).

FERMIONS

The elementary (indivisible)
particles that make up matter
fall into two groups: leptons
and quarks. Only a few of
each family are widespread
in today's universe.

The subatomic world
Using particle accelerators (see p.121)
to break apart atoms and create
short-lived and unstable particles,
physicists have assembled the so-called
Standard Model (see pp.124–125) of
particle physics.

FERMIONS

ELEMENTARY FERMIONS

QUARKS
- UP
- DOWN
- CHARM
- STRANGE
- TOP
- BOTTOM

LEPTONS
- ELECTRON
- ELECTRON NEUTRINO
- MUON
- MUON NEUTRINO
- TAU PARTICLE
- TAU NEUTRINO

SUBATOMIC PARTICLES

HADRONS

These composite particles are made of quarks. Baryons are themselves fermionic in nature, while mesons behave as bosons (see p.68)

COMPOSITE PARTICLES

BARYONS

- PROTON
- NEUTRON
- LAMBDA PARTICLE
- OTHERS

MESONS

- POSITIVE PION
- NEGATIVE KAON
- OTHERS

BOSONS

ELEMENTARY BOSONS

- PHOTON
- GLUON
- W- BOSON
- W+ BOSON
- Z BOSON
- HIGGS BOSON

BOSONS

Elementary bosons mostly act as "messengers," transmitting forces between particles of matter. They behave very differently from fermions.

WHAT IS LIGHT?

Particles with electric charge can exchange forces between each other by emitting electromagnetic radiation. These moving waves consist of oscillating electric and magnetic fields aligned at right angles so that changes in one reinforce the other. The properties and effects of these waves are determined by the frequency at which they repeat, giving rise to different forms of radiation such as X-rays, radio waves, and visible light.

THE SPECTRUM
Scientists classify electromagnetic waves according to their wavelength and frequency, from long-wavelength, low-frequency radio waves, through microwaves, infrared, and visible light, to shorter wavelength and higher frequency ultraviolet, X-rays, and gamma rays.

Self-propagating wave
As the electric and magnetic fields in an electromagnetic wave vary, they reinforce each other, allowing the wave to travel for very long distances.

DIRECTION OF MOTION

ELECTRIC FIELD

MAGNETIC FIELD

OSCILLATING FIELDS
Light consists of electrical and magnetic waves that are at right angles to each other and perpendicular to its direction of motion.

RADIO WAVES   MICROWAVES   INFRARED   VISIBLE LIGHT   UV   X-RAYS   GAMMA RAYS

ELECTROMAGNETIC SPECTRUM

QUANTUM CONSTANT

At the level of individual atoms and subatomic particles, electrically charged matter emits and absorbs electromagnetic radiation as small packets called photons. These tiny objects are "quanta" (discrete, self-contained bundles) of electromagnetic energy. The amount of energy a photon contains is determined by a simple calculation involving its frequency, the speed of light, and a number known as Planck's constant.

$$E = h \nu$$

FREQUENCY

The frequency of a photon is the number of waves that would pass a point in a single second if the wave was continuous—as the photon's wavelength shrinks, its frequency increases.

PHOTON'S ENERGY

The energy carried by a single photon is greater for higher frequencies and shorter wavelengths of light.

PLANCK'S CONSTANT

This fundamental constant of quantum physics defines the relationship between a photon's energy and its frequency or wavelength. It means that energy is only delivered in discrete units.

$$h = 6.62607015 \times 10^{-34}$$

JOULE SECONDS

PLANCK'S CONSTANT

RIPPLES IN SPACE

Wavelike behavior is fundamental not just to electromagnetic radiation (see p.14) but also to the quantum behavior of particles. Unlike particles, waves can pass through each other to boost the overall disturbance in some places and decrease it in others (an effect called interference) and also spread into the "shadows" cast by barriers (diffraction). When they encounter a boundary between two different materials, waves can be bounced back (reflection) or slowed down and deflected onto new paths (refraction).

ENERGY TRANSFERENCE

Waves are repeated oscillations (fluctuations) around a fixed midpoint. While waves transfer energy, they do not carry matter from one place to another.

AMPLITUDE

The amplitude of a wave is the maximum displacement a field or particle oscillates from its central equilibrium position.

Wave essentials

A wave's frequency is the number of times it oscillates per second, while wavelength is the distance covered by one complete oscillation.

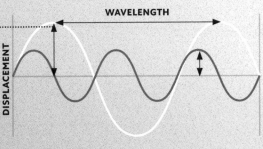

WAVELENGTH

DISPLACEMENT

Interference

When the crests of two waves of the same frequency line up, they form a wave with greater amplitude (called constructive interference). Destructive interference occurs when the troughs of one wave partially or entirely cancel out the peaks of another.

COMBINED WAVE

WAVES CANCEL OUT

WAVE OR PARTICLE

At quantum scales, the dividing line between particles and waves (see pp.16–17) becomes blurred, with strange results. It is possible to design experiments that detect individual particlelike packets of energy, such as photons (quanta of electromagnetic radiation; see p.14), and at the same time demonstrate their wavelike behavior. Photons may arrive one at a time at a detector on the opposite side of two small slits, yet the pattern they build up can only be explained by each photon deciding on its location based upon wavelike interference (see pp.16–17).

Double slit experiment

A famous experiment conducted in 1800 to prove the wave nature of light can be adapted to show the wavelike nature of electrons and other particles.

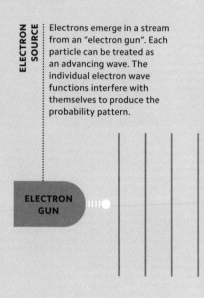

ELECTRON SOURCE

Electrons emerge in a stream from an "electron gun". Each particle can be treated as an advancing wave. The individual electron wave functions interfere with themselves to produce the probability pattern.

ELECTRON GUN

The effect of measurements

One of the strangest aspects of quantum theory is that wave or particle behavior can be determined by the process of measurement.

PARTICLE DETECTOR

If we measure which slit each photon or electron passes through, they behave as particles at that slit and lose the wavelike behavior that existed prior to the slits.

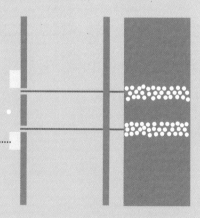

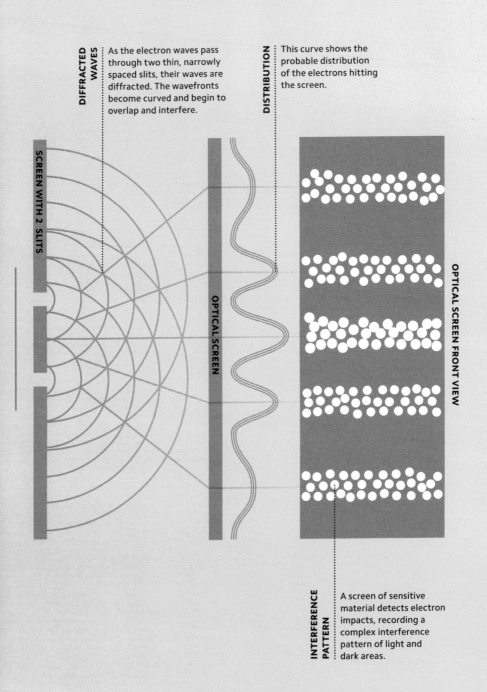

DIFFRACTED WAVES As the electron waves pass through two thin, narrowly spaced slits, their waves are diffracted. The wavefronts become curved and begin to overlap and interfere.

DISTRIBUTION This curve shows the probable distribution of the electrons hitting the screen.

SCREEN WITH 2 SLITS

OPTICAL SCREEN

OPTICAL SCREEN FRONT VIEW

INTERFERENCE PATTERN A screen of sensitive material detects electron impacts, recording a complex interference pattern of light and dark areas.

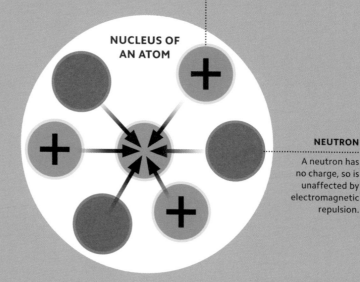

Nuclear force
The nuclear force binds quarks (and protons and neutrons) together, opposing electromagnetic repulsion (see p.22).

PROTON

Protons repel each other, but this repulsion is overcome by the stronger nuclear force.

NUCLEUS OF AN ATOM

NEUTRON

A neutron has no charge, so is unaffected by electromagnetic repulsion.

HOLDING IT TOGETHER

Four fundamental forces are responsible for binding the matter particles in the universe together, and each is governed by quantum physics to some extent. The most powerful of these forces, known as the strong force, only works on tiny scales of about one million-billionth of a yard. This force bonds quark particles together to form protons and neutrons, and produces a nuclear force that binds these to form atomic nuclei. The strong force is carried by particles called gluons.

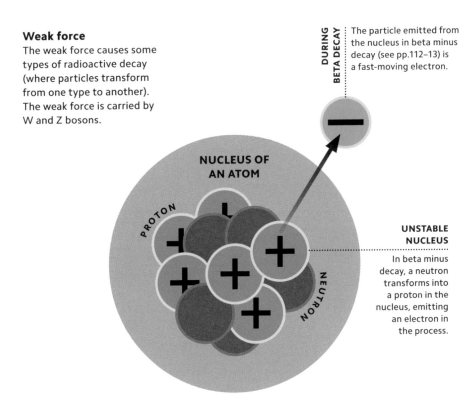

Weak force
The weak force causes some types of radioactive decay (where particles transform from one type to another). The weak force is carried by W and Z bosons.

DURING BETA DECAY The particle emitted from the nucleus in beta minus decay (see pp.112–13) is a fast-moving electron.

NUCLEUS OF AN ATOM

PROTON

NEUTRON

UNSTABLE NUCLEUS
In beta minus decay, a neutron transforms into a proton in the nucleus, emitting an electron in the process.

THE FORCE OF DECAY

The weak force, as its name suggests, is less powerful than the strong and electromagnetic forces, and it operates over even smaller scales, only making itself felt at ranges below the diameter of a proton. However, weak interactions are hugely important as they can influence matter particles of all types (both quarks and leptons), and the weak force is the only one of the fundamental forces that can turn one type of particle into another type.

Infinite range

On the atomic scale, electromagnetism is the attractive force between protons and electrons. Electromagnetic radiation is carried by massless particles called photons (see p.14).

Electrons and the protons in the nucleus are attracted to each other, keeping them together in the atom.

HELD IN ORBIT

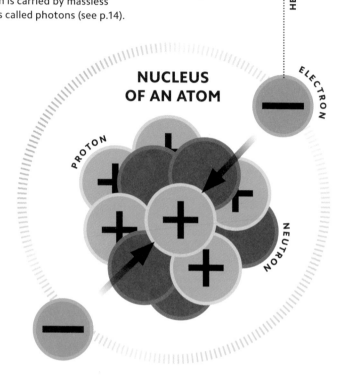

NUCLEUS
OF AN ATOM

ELECTRON

PROTON

NEUTRON

OPPOSITES ATTRACT

It is the force of electromagnetism that attracts particles of opposite electric charge and also repels particles with the same electric charge. Electromagnetism has an infinite range, not only binding atoms together but also shining as light across vast distances in the cosmos, although its strength decreases rapidly with distance.

DRAWN TOGETHER

Gravitation is an attractive force between objects with mass. It is extremely weak and only becomes apparent between objects of large mass, yet, like electromagnetism, it has an infinite range. The best model for understanding gravitation is Einstein's General Relativity, a theory that seems completely separate from quantum physics. Understanding how gravity works on the level of particles poses many baffling questions (see pp.136–45).

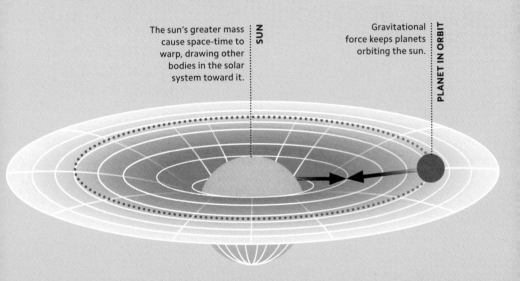

The sun's greater mass cause space-time to warp, drawing other bodies in the solar system toward it.

SUN

Gravitational force keeps planets orbiting the sun.

PLANET IN ORBIT

Space-time
Einstein described the three dimensions of space and the dimension of time as a four-dimensional grid called space-time. General Relativity explains gravity as arising from distortions in space-time by massive objects.

PRE-QU
PUZZLE

ANTUM
S

Quantum physics began as an attempt by scientists to explain a number of apparently separate puzzles in early 20th-century physics. These puzzles affected the nature of light emitted by objects heated to different temperatures, the internal structure of the atom, and the interaction between light and matter. Together, they led to the realization that electromagnetic waves are emitted and delivered in small, discrete packets of energy known as photons, and hinted at deeper mysteries in the behavior of subatomic particles.

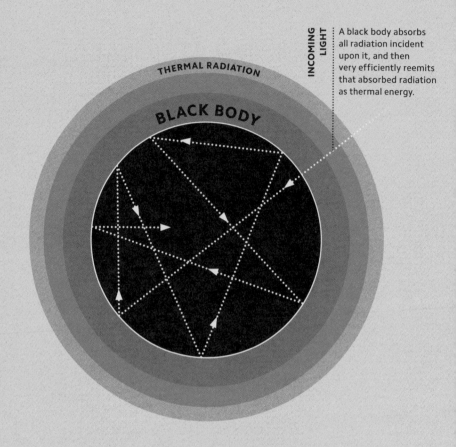

THERMAL RADIATION

BLACK BODY

INCOMING LIGHT

A black body absorbs all radiation incident upon it, and then very efficiently reemits that absorbed radiation as thermal energy.

IDEAL BODIES

In order to understand how objects emit electromagnetic radiation when they are heated, scientists use an idealized object called a "black body." All but the coldest objects emit some form of radiation, but because most will also reflect radiation from their surroundings, it can be hard to measure how much radiation is actually being released. A black body has a pitch-black, completely nonreflective surface whose radiation is dependent only on its temperature.

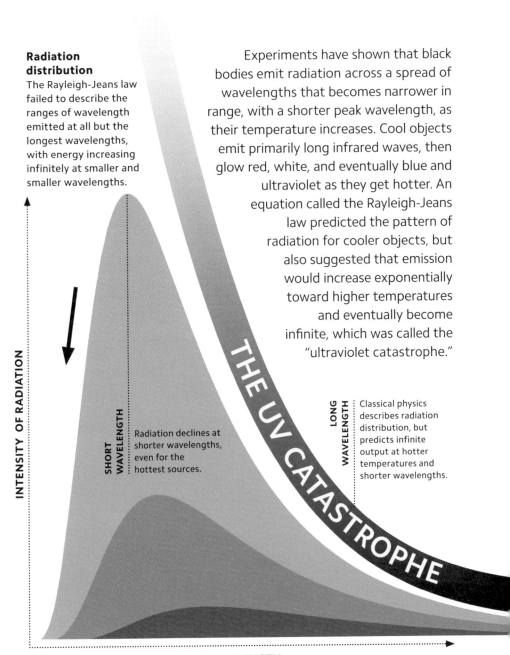

Radiation distribution
The Rayleigh-Jeans law failed to describe the ranges of wavelength emitted at all but the longest wavelengths, with energy increasing infinitely at smaller and smaller wavelengths.

Experiments have shown that black bodies emit radiation across a spread of wavelengths that becomes narrower in range, with a shorter peak wavelength, as their temperature increases. Cool objects emit primarily long infrared waves, then glow red, white, and eventually blue and ultraviolet as they get hotter. An equation called the Rayleigh-Jeans law predicted the pattern of radiation for cooler objects, but also suggested that emission would increase exponentially toward higher temperatures and eventually become infinite, which was called the "ultraviolet catastrophe."

THE UV CATASTROPHE

INTENSITY OF RADIATION

SHORT WAVELENGTH

Radiation declines at shorter wavelengths, even for the hottest sources.

LONG WAVELENGTH

Classical physics describes radiation distribution, but predicts infinite output at hotter temperatures and shorter wavelengths.

WAVELENGTH

PACKETS OF ENERGY

In 1900, Max Planck showed a way to avoid the ultraviolet catastrophe (see p.27) and make the theoretical emissions of black bodies (see p.26) match with their measured behavior. What if energy was being released not in a continuous stream, but as small, discrete bursts (or packets of energy), each with a distinct wavelength? Planck called these bursts "light quanta," and assumed that their production had something to do with the emission process rather than being a property of light itself (see p.14).

GRADUAL INCREASE

Classical physics
In the pre-quantum view, properties of particles, such as the energy they hold, vary continuously and may have any value.

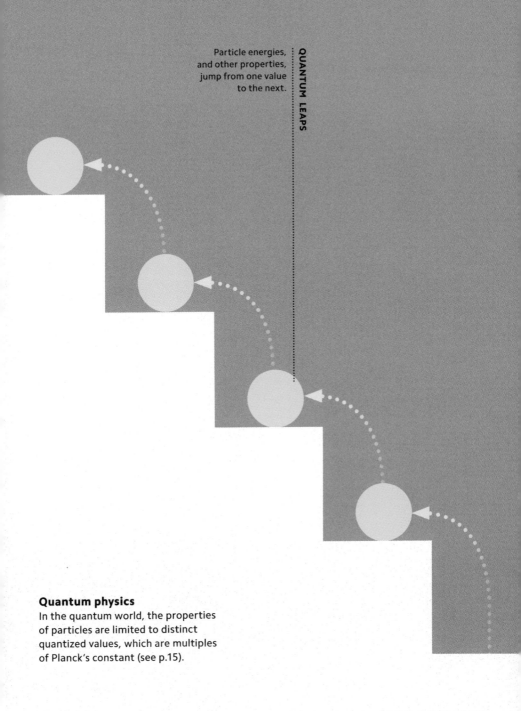

Particle energies, and other properties, jump from one value to the next.

QUANTUM LEAPS

Quantum physics

In the quantum world, the properties of particles are limited to distinct quantized values, which are multiples of Planck's constant (see p.15).

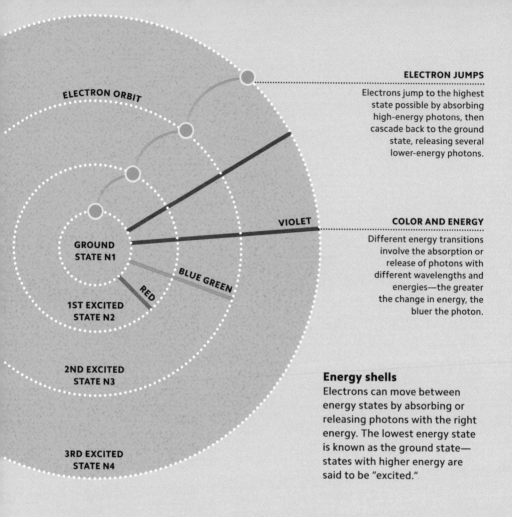

ELECTRON ORBIT

ELECTRON JUMPS
Electrons jump to the highest
state possible by absorbing
high-energy photons, then
cascade back to the ground
state, releasing several
lower-energy photons.

VIOLET

COLOR AND ENERGY
Different energy transitions
involve the absorption or
release of photons with
different wavelengths and
energies—the greater
the change in energy, the
bluer the photon.

**GROUND
STATE N1**

BLUE GREEN

RED

**1ST EXCITED
STATE N2**

**2ND EXCITED
STATE N3**

Energy shells
Electrons can move between
energy states by absorbing or
releasing photons with the right
energy. The lowest energy state
is known as the ground state—
states with higher energy are
said to be "excited."

**3RD EXCITED
STATE N4**

ENERGETIC STATES

Early 20th-century physicists wrestled with how the structure
of atoms (see pp.10–11) related to the way they emitted or
absorbed radiation. In 1913, Niels Bohr proposed a model in
which electrons orbited in shells at various distances from the
nucleus, giving each a distinctive energy state. Atoms absorbed
or emitted quanta of electromagnetic energy whose wavelengths
corresponded to the difference between these states.

CLOUDS OF PROBABILITIES

Discoveries in the 1920s revealed that atomic structure is more complex than the simple Bohr model. The modern model shows that electrons occupy a series of "orbitals"—shells and subshells with a variety of shapes. As it is impossible to know all of their properties at a single instant (see pp.42–43), it is more accurate to think of these orbitals as fuzzy regions where the electrons are likely to be found—for some purposes, the electron's properties are effectively "smeared out" across the orbital.

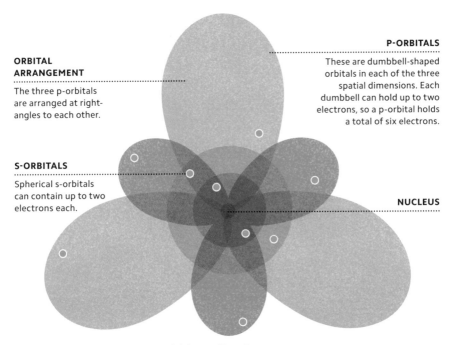

ORBITAL ARRANGEMENT

The three p-orbitals are arranged at right-angles to each other.

S-ORBITALS

Spherical s-orbitals can contain up to two electrons each.

P-ORBITALS

These are dumbbell-shaped orbitals in each of the three spatial dimensions. Each dumbbell can hold up to two electrons, so a p-orbital holds a total of six electrons.

NUCLEUS

Orbiting a fluorine atom
A fluorine atom contains nine electrons, two each in its inner two S-orbitals and five in the first p-orbital.

Red light
Each photon of red light does not possess enough energy to liberate individual electrons. Making the light brighter just increases the number of low-energy photons.

LOW ENERGY

METAL SURFACE

PHOTON ENERGY

The photoelectric effect causes electric current to flow from the surface of certain metals when they are bombarded with light. However, it only works when that light is shorter than a certain wavelength; even intense bombardment with longer-wavelength light cannot trigger it. In 1905, Albert Einstein determined that this was because the effect depends on electrons being struck by individual light quanta. Based on his discovery, Einstein argued that all radiation takes the form of quanta or "photons."

Albert Einstein won his only Nobel Prize for describing the photoelectric effect, not for his theories of relativity.

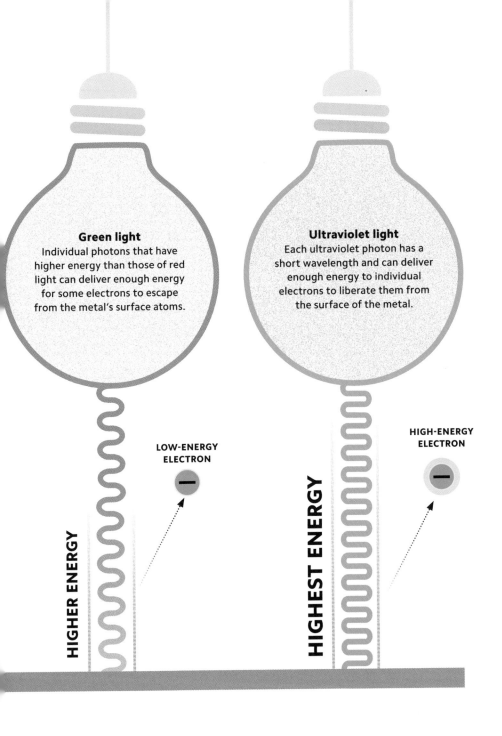

Green light
Individual photons that have higher energy than those of red light can deliver enough energy for some electrons to escape from the metal's surface atoms.

Ultraviolet light
Each ultraviolet photon has a short wavelength and can deliver enough energy to individual electrons to liberate them from the surface of the metal.

LOW-ENERGY
ELECTRON

HIGH-ENERGY
ELECTRON

HIGHER ENERGY

HIGHEST ENERGY

THE W
FUNC

A V E
T I O N

in classical physics, the nature of a system at any point in time can be precisely calculated using deterministic rules, such as Isaac Newton's laws of mechanics. In the quantum world, however, systems unravel unpredictably. A quantum system is best described with mathematical "wave function," which gives the probability of finding it in a certain state at a certain time. Quantum systems that could be in one of several states can be described with a superposition of all these possible states, although this superposition always "collapses" into a single state when a measurement is taken. It is this collapse of the wave function that creates unpredictability.

Basic wave function
This image is an example of a wave function for a particle moving in one dimension.

At the greatest amplitude—the distance from the central equilibrium point to the peak—is the point where there is the highest probability of finding the particle.

HIGH PROBABILITY

DESCRIBING A QUANTUM STATE

Objects on the quantum scale behave in unpredictable ways; for instance, it is impossible to calculate with certainty a particle's state at a given time. Instead, its state is described mathematically with a wave function that varies in space and time. The probability that the particle will be found at a certain place and time is related to the amplitude of the wave function multiplied by itself (see p.40).

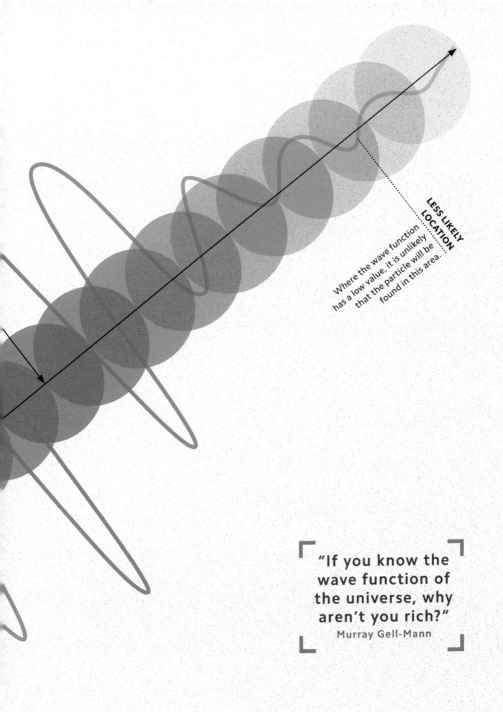

LESS LIKELY
LOCATION

Where the wave function has a low value, it is unlikely that the particle will be found in this area.

"If you know the wave function of the universe, why aren't you rich?"
Murray Gell-Mann

> "Causality applies only to a system which is left undisturbed."
> Paul Dirac

SPIN STATES
....
An electron can exist in a superposition of different states, such as spin up or spin down.

IN TWO PLACES AT ONCE

In classical physics, waves can be added together to form another wave (superposition). Similarly, quantum states—described by wave functions—can be combined to form another quantum state. This is known as quantum superposition. A quantum system that could be found in one of multiple states (e.g. an electron could have spin up or spin down, see p.66) can be described with a superposition of all these possible states.

PARTICLE POSITIONS

An unobserved particle can be conceived as existing in every possible position at the same time.

Supporting waves

Any two quantum states can be added together to create another valid quantum state. When the waves are identical, as in this example, they reinforce one another in superposition.

ELECTRON

ELECTRON

SQUARED
WAVE

The Born rule, named after the German physicist Max Born, is used to calculate the probability of finding a system in a certain state, based on the wave function (see pp.36–37) that is used to describe the system. For any enclosed system, the probability of finding a particle in a certain location is proportional to the square of the wave function's magnitude at that location. The wave function, and therefore the probability, is dependent upon the energy of a particle (see p.30).

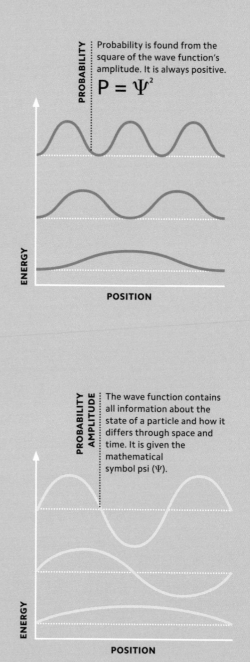

PROBABILITY

Probability is found from the square of the wave function's amplitude. It is always positive.

$$P = \Psi^2$$

ENERGY

POSITION

PROBABILITY AMPLITUDE

The wave function contains all information about the state of a particle and how it differs through space and time. It is given the mathematical symbol psi (Ψ).

ENERGY

POSITION

WAVE TRANSFORMATIONS

Fourier transforms are mathematical operations that represent any function as a composition of basic waves with various frequencies. This allows for switching between "domains" such as time and frequency. In quantum mechanics, they are used to switch between position and momentum. This means that a quantum state represented by a wave function of position can be switched to being represented by a wave function of momentum, and vice versa.

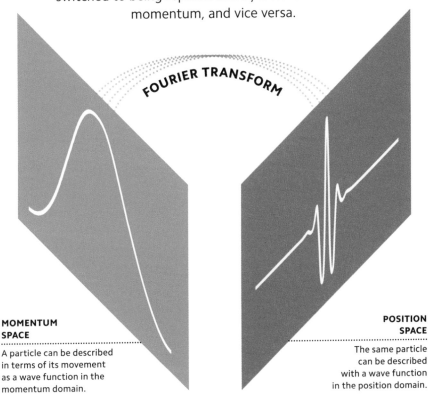

FOURIER TRANSFORM

MOMENTUM SPACE

A particle can be described in terms of its movement as a wave function in the momentum domain.

POSITION SPACE

The same particle can be described with a wave function in the position domain.

NOT ALL IS KNOWABLE

Unlike in classical mechanics, in the quantum world some pairs of physical quantities cannot be calculated with certainty. For example, it is impossible to know the exact position and momentum of a particle at the same time; the more precisely one of these quantities is determined, the less precisely the other can be determined. This is known as the uncertainty principle or Heisenberg's uncertainty principle, after German physicist Werner Heisenberg, who discovered the law along with his early framework of quantum mechanics (known as matrix mechanics).

POSITION

The more certain the position of a particle becomes, the less is known about its momentum.

"We can never know anything."
Werner Heisenberg

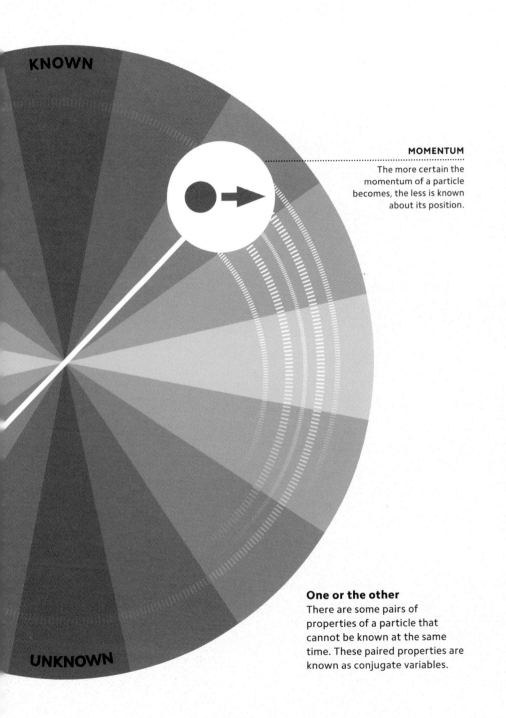

KNOWN

MOMENTUM
The more certain the momentum of a particle becomes, the less is known about its position.

UNKNOWN

One or the other
There are some pairs of properties of a particle that cannot be known at the same time. These paired properties are known as conjugate variables.

This part of the equation gives the rate of change of the wave function with respect to time.

The equation uses complex numbers that contain the imaginary number "i," which is equivalent to the square root of minus one ($\sqrt{-1}$).

The wave function is represented by the Greek letter psi.

$$i\hbar\frac{\partial}{\partial t}\Psi = \hat{H}\Psi$$

The reduced Planck constant is the quantum of variables such as spin, momentum, energy, space, and time.

The Hamiltonian is a function that, when applied to the wave function, represents the sum of kinetic and potential energies for all particles described by the wave function.

Time dependent

Schrödinger's equation can be written in many different forms. The time-dependent equation shown here describes a system evolving as time passes.

PREDICTING CHANGE

The Schrödinger equation
determines the evolution of a
wave function (se pp.36–37),
predicting the future behavior of
the system of superposed states it
describes. It can be considered the
quantum equivalent of Newton's laws
of motion, which predict changes
to a classical system over time. The
future behavior of a system cannot be
determined with certainty, but the
equation allows us to calculate the
probability of finding a system in a
certain state at some later time.

"Where did we get that
[Schrödinger's equation] from?
It's not possible to derive it
from anything you know.
It came out of the mind
of Schrödinger."
Richard Feynman

MEASUREMENT TAKEN

MEASURING POSITION

Some measurement can influence
what is being measured, such as
when a photon is used to measure
the position of an electron.

PHOTON

EVADING MEASUREMENT

The Schrödinger equation (see pp.44–45)
describes the evolution of a wave function—
giving the probability of finding the system in
various states in a superposition at any given time—but
when the system is measured, it is always found in a single
state. It is impossible to observe the wave function of superposed
states and say which will be seen when the system is measured; this
mystery has led to different interpretations of quantum mechanics.

AFTER MEASUREMENT

PHOTON
The interaction with
the electron has also
altered the photon.

ELECTRON MOVES
Having absorbed energy from
the photon, the electron's wave
function has changed because of
changes to its energy and position.

"The problem of
measurement ... is the
problem of where the
measurement begins
and ends, and where the
observer begins and ends."
John Stewart Bell

INSTANT COLLAPSE

When a quantum system is measured, the wave function (see pp.36–37) representing the superposition of states, with probabilities assigned to different states, stops evolving and is reduced to a single definite state. This process is known as wave function collapse. The evolution of the wave function is determined by the Schrödinger wave equation (see pp.44–45); at any specific point in time, wave function collapse leaves only one possible outcome.

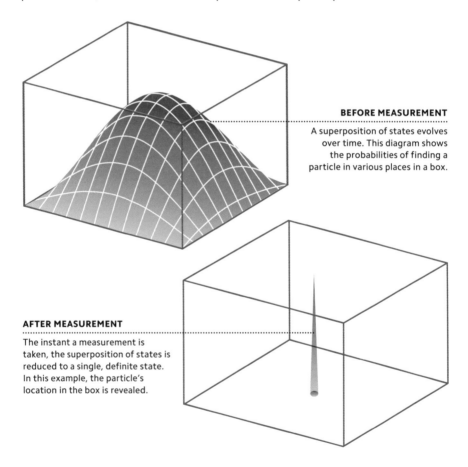

BEFORE MEASUREMENT

A superposition of states evolves over time. This diagram shows the probabilities of finding a particle in various places in a box.

AFTER MEASUREMENT

The instant a measurement is taken, the superposition of states is reduced to a single, definite state. In this example, the particle's location in the box is revealed.

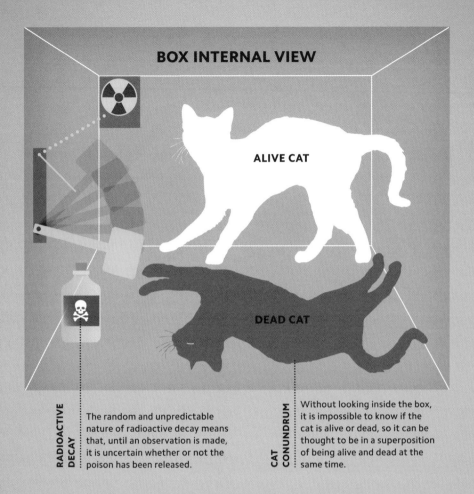

ALIVE CAT

DEAD CAT

RADIOACTIVE DECAY
The random and unpredictable nature of radioactive decay means that, until an observation is made, it is uncertain whether or not the poison has been released.

CAT CONUNDRUM
Without looking inside the box, it is impossible to know if the cat is alive or dead, so it can be thought to be in a superposition of being alive and dead at the same time.

PARADOX IN A BOX

The mystery of what occurs before wave function collapse inspired a famous thought experiment known as "Schrödinger's cat." In this thought experiment, a cat in a box could be killed at any time by poison released via nuclear decay. In one interpretation of quantum mechanics (see pp.52–53), the superposed states of dead and alive exist until an observer opens the box. Erwin Schrödinger devised this thought experiment to highlight the absurdity of a cat being both dead and alive until someone checks on it.

INTERPR
OF QUANT
MECHANI

ETATIONS
UM
CS

The wavelike evolution and unpredictable wave function collapse of quantum physics is hard to reconcile with more familiar classical physics in which outcomes, positions, and behaviors are well defined. In order to explain why quantum uncertainty is not seen in the everyday world, scientists and philosophers have devised many different "interpretations" of quantum physics. Some attempt to simply get rid of uncertainty above a certain scale by forcing the wave function to collapse, while others take more ingenious routes to explain why we never come across uncollapsed wave functions "in the wild."

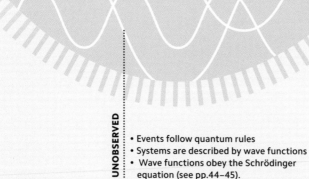

UNOBSERVED

- Events follow quantum rules
- Systems are described by wave functions
- Wave functions obey the Schrödinger equation (see pp.44–45).

"I think that a particle must have a separate reality independent of measurements. ... I like to think the moon is there even if I am not looking at it."
Albert Einstein

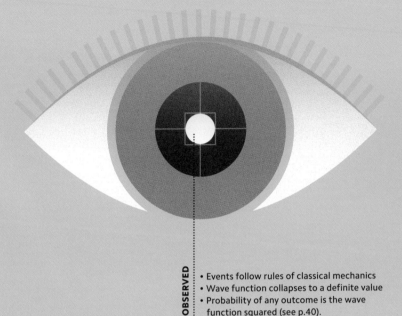

- Events follow rules of classical mechanics
- Wave function collapses to a definite value
- Probability of any outcome is the wave function squared (see p.40).

UNDERSTANDING QUANTUM PHYSICS

The Copenhagen interpretation was the earliest attempt to bridge the gap between the quantum world and classical physics. In this view, the act of an observer measuring the state of a quantum system causes it to resolve instantaneously to a single value—a phenomenon called the collapse of the wave function (see p.48). The wave equation itself is treated as merely offering a measure of the probability that different values will be detected when the observation takes place (see pp.36–37).

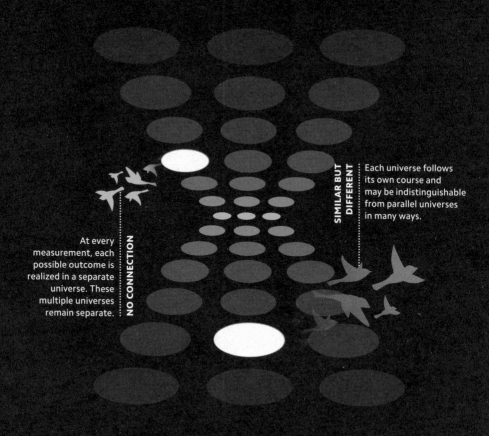

SIMILAR BUT DIFFERENT

Each universe follows its own course and may be indistinguishable from parallel universes in many ways.

At every measurement, each possible outcome is realized in a separate universe. These multiple universes remain separate.

NO CONNECTION

EVERYTHING CAN AND DOES HAPPEN

This quantum interpretation, developed by physicist Hugh Everett III, considers the wave function to be a particle's true nature, and wave function collapse (see p.48) to be impossible. Instead, measurement creates a multitude of parallel universes—one for each possible outcome of the measurement process. The wave function remains uncollapsed as a whole, but an observer ends up in one of these parallel universes where it appears to have collapsed.

INFINITE REPETITION

The cosmological interpretation is one way of understanding the parallel realities of the many worlds theory. It suggests that our universe has a truly infinite extent in space, within which every event is repeated an infinite number of times. The wave function links these remote events, so that while we may observe one outcome of a measurement at our point in space, elsewhere other versions of us are making the same observations with different results.

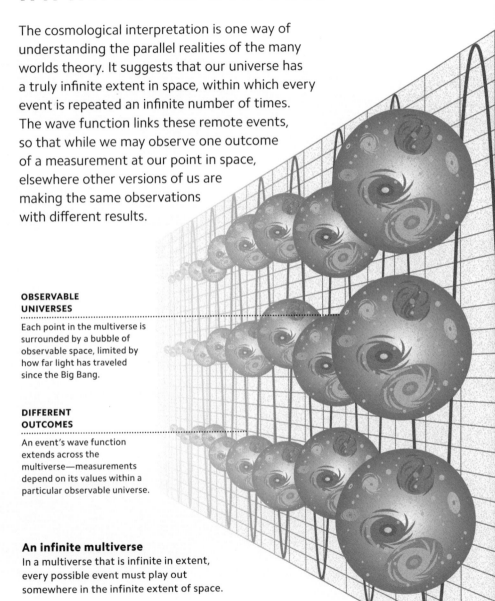

OBSERVABLE UNIVERSES

Each point in the multiverse is surrounded by a bubble of observable space, limited by how far light has traveled since the Big Bang.

DIFFERENT OUTCOMES

An event's wave function extends across the multiverse—measurements depend on its values within a particular observable universe.

An infinite multiverse

In a multiverse that is infinite in extent, every possible event must play out somewhere in the infinite extent of space.

In pilot wave theory, results of
quantum measurements are
guided by undetectable
matter waves steering
collections of particles.

UNSEEN INFLUENCE

The Copenhagen interpretation
assumes that entangled particles (see
pp.52–53) somehow share information
instantaneously. Hidden-variable
interpretations suggest one way to avoid this
apparently faster-than-light transfer of
information—a particle has undetected
properties that guide the wave function's
collapse. Pilot wave theories propose
the existence of unseen quantum waves
that do a similar job of "steering"
wave functions to collapse
toward certain properties.

"We are not sufficiently astonished by the
fact that any science may be possible."
Louis de Broglie

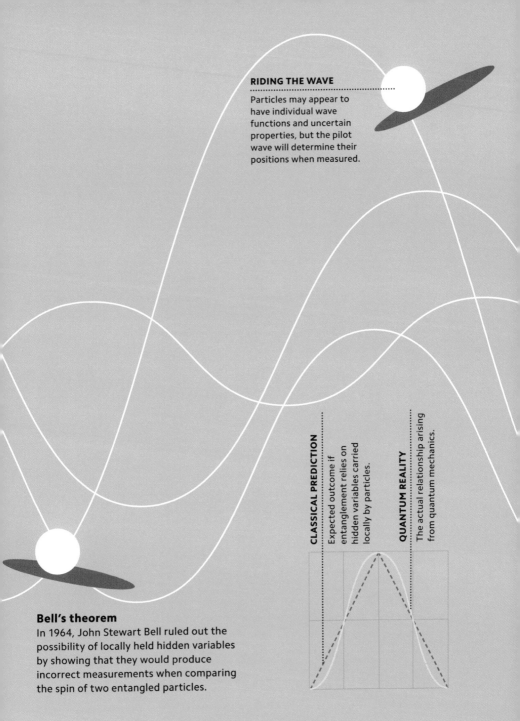

RIDING THE WAVE

Particles may appear to have individual wave functions and uncertain properties, but the pilot wave will determine their positions when measured.

CLASSICAL PREDICTION

Expected outcome if entanglement relies on hidden variables carried locally by particles.

QUANTUM REALITY

The actual relationship arising from quantum mechanics.

Bell's theorem

In 1964, John Stewart Bell ruled out the possibility of locally held hidden variables by showing that they would produce incorrect measurements when comparing the spin of two entangled particles.

Interference between waves moving forward and backward in time produces an appearance of instantaneous wave-function collapse.

A QUANTUM HANDSHAKE

The transactional interpretation offers a possible explanation for the way that interactions between quantum systems actually take place. It suggests that the wave function generates waves that move out into the surroundings and both forward and backward in time. Other systems (including observers) generate similar waves. When the forward-moving wave from a quantum system encounters a backward-moving wave from another source, the "handshake" process that occurs resolves the properties of the quantum system.

OBSERVER

Chance encounter
The measured position, momentum, and other properties of a quantum particle are determined by the interaction between its wave function and the observer's wave.

Wave function half-life
Though initially uncertain, a wave function has a certain probability of collapsing spontaneously, somewhat like the half-life of radioactive decay. Collapse may be accelerated by interactions with the collapsing wave functions of other particles.

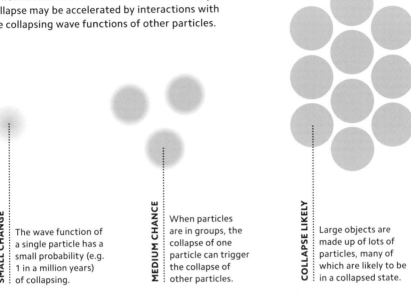

SMALL CHANGE
The wave function of a single particle has a small probability (e.g. 1 in a million years) of collapsing.

MEDIUM CHANCE
When particles are in groups, the collapse of one particle can trigger the collapse of other particles.

COLLAPSE LIKELY
Large objects are made up of lots of particles, many of which are likely to be in a collapsed state.

SPONTANEOUS COLLAPSE

The Copenhagen interpretation's suggestion that measurement or observation trigger the collapse of the wave function raises the troubling question: what happens when no measurements are taken? Alternative or spontaneous collapse theories get around this by suggesting that the collapse happens automatically and randomly, without a trigger from the outside world. A wave function's likelihood of collapse can still be influenced by the wave functions of particles surrounding it.

PARTICLE WILDERNESS

The less interaction a quantum particle has with its environment, the wider the range of its possible states. Due to decoherence—the entangling of the particle state and that of the world around it— eventually there will be only one outcome.

Eventually one wave function proves itself fittest for the environment—it is this one that will determine the results of classical measurements.

LONE SURVIVOR

SURVIVAL OF THE FITTEST

If hidden variables (see pp.56–57) do not guide the wave function to collapse into a certain state, perhaps something else does? Quantum Darwinism, as its name suggests, involves an idea similar to Charles Darwin's evolutionary "survival of the fittest." According to this interpretation, interactions between a particle and factors in its environment gradually filter the possible end states of its wave function until it settles on a single outcome known as a "pointer state."

THE OBSERVER'S BELIEF

Named after 18th-century statistician Thomas Bayes, and also known as QBism (pronounced "cubism"), this interpretation of quantum physics puts the observer at the center of the theory. In the same way that Bayesian statistics allow people to adjust their ideas about the likelihood of events as more information becomes available, so according to QBism, the wave function is merely a representation of the observer's subjective information and beliefs about possible different outcomes.

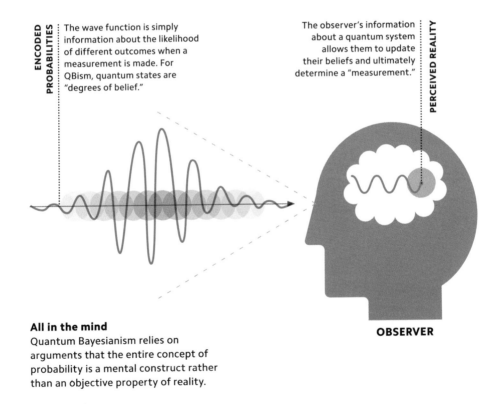

ENCODED PROBABILITIES

The wave function is simply information about the likelihood of different outcomes when a measurement is made. For QBism, quantum states are "degrees of belief."

PERCEIVED REALITY

The observer's information about a quantum system allows them to update their beliefs and ultimately determine a "measurement."

OBSERVER

All in the mind
Quantum Bayesianism relies on arguments that the entire concept of probability is a mental construct rather than an objective property of reality.

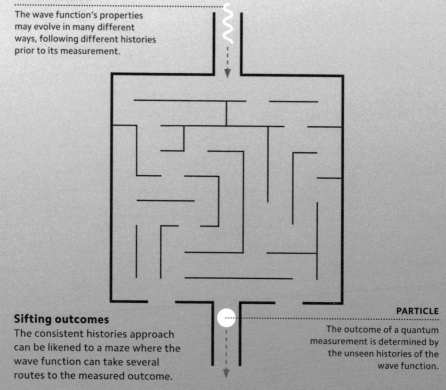

The wave function's properties may evolve in many different ways, following different histories prior to its measurement.

Sifting outcomes

The consistent histories approach can be likened to a maze where the wave function can take several routes to the measured outcome.

PARTICLE

The outcome of a quantum measurement is determined by the unseen histories of the wave function.

THROUGH THE MAZE

Sometimes referred to as "Copenhagen done right," the consistent histories approach attempts to marry quantum mechanics with the "classical" rules of mathematical probability. In this approach, the wave function never collapses, but measurement of the quantum system reveals properties related to the state of the wave function at that time. A set of potential "histories" describing the possible evolution of the system can be mapped out, with a certain probability assigned to each.

DIFFERENT VIEWS

Albert Einstein's famous theory of Special Relativity explains how observers in different "frames of reference" (for instance, moving at different speeds) can differ in their interpretation of events. The relational interpretation applies the same idea to the wave function, suggesting that it may simultaneously display different states of collapse (or remain uncollapsed) for different observers depending on their location, motion, and other properties; the quantum system and its observers are, in fact, described together by a combined wave function.

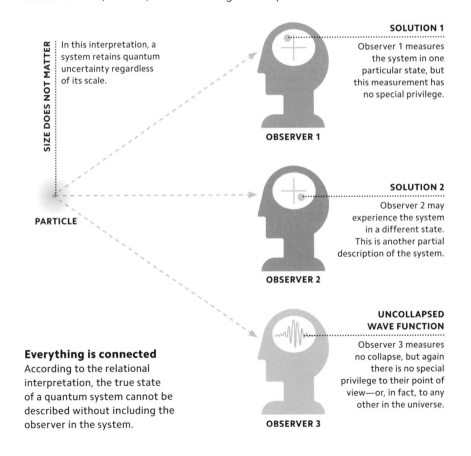

SIZE DOES NOT MATTER

In this interpretation, a system retains quantum uncertainty regardless of its scale.

PARTICLE

SOLUTION 1

Observer 1 measures the system in one particular state, but this measurement has no special privilege.

OBSERVER 1

SOLUTION 2

Observer 2 may experience the system in a different state. This is another partial description of the system.

OBSERVER 2

UNCOLLAPSED WAVE FUNCTION

Observer 3 measures no collapse, but again there is no special privilege to their point of view—or, in fact, to any other in the universe.

OBSERVER 3

Everything is connected
According to the relational interpretation, the true state of a quantum system cannot be described without including the observer in the system.

QUANT
PHENO

U M
M E N A

When put into practice, the strange rules of quantum physics produces a wide variety of unusual phenomena. The quantum revolution of the 1920s helped to resolve some of the biggest outstanding mysteries in the physics of the time (such as what forces governed the internal structure of atoms, and how some unstable atomic nuclei decay). However, it also made predictions of strange behaviors and unusual states of matter that troubled many physicists at the time, but which were demonstrated through ingenious experiments later in the 20th century.

"SPIN"

In large-scale physics, spinning objects naturally possess a form of momentum due to a combination of their mass and rotation. Quantum particles possess a property (called intrinsic angular momentum, or spin) that was initially thought to be them spinning on the spot, but later turned out to be something much more curious. Spin does not refer to an actual physical rotation of particles, but shares many of the same characteristics as classical rotation and produces some similar effects.

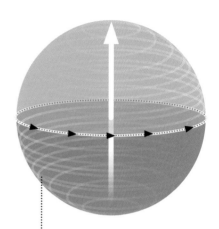

SPIN UP The spin of a particle is measured against one of three dimensions, most often the dimension defined as the z-axis. When cast on this axis, the spin points in one of two directions: positive spin up or negative spin down.

SPIN DOWN Rotating a "spin-up" +½ electron through 360° results in its spin being reversed to "spin-down" or -½.

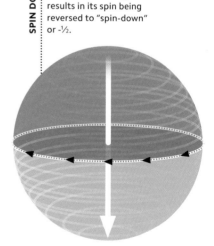

> "But then the discovery of electron spin changed this picture considerably. The electron was not symmetrical. ... They are not simple, not so elementary as we had thought before."
> Werner Heisenberg

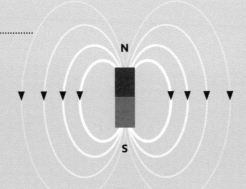

BAGNETIC DIPOLES
....
In this familiar bar
magnet, countless
dipoles (molecules
with a positively and
a negatively charged
side) are aligned
to produce a strong
magnetic field.

Field lines
Magnetic fields
are often depicted
using lines that
indicate the
strength and
direction of
their influence.

FIELDS OF ATTRACTION

While electric charge is a fundamental property of many
particles, magnetism is not—instead it is an effect that arises
from electric charges in motion. Charged subatomic particles
such as electrons and quarks develop their own weak magnetic
fields, known as magnetic moments, as a consequence of their
intrinsic angular momentum, or spin. Like spin and charge
themselves, magnetic moments are quantized—only capable
of taking on certain discrete values.

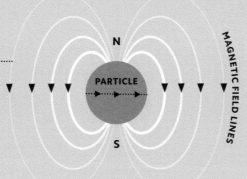

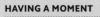
HAVING A MOMENT
An effect called
electromagnetic
induction means
that any spinning
electrically charged
particle induces
(creates) a magnetic
field similar to that
of a bar magnet.

THE GREAT DIVIDE

The property of spin defines a fundamental division between elementary subatomic particles. Particles associated with the structure of matter all have spin values of either $+\frac{1}{2}$ or $-\frac{1}{2}$, while those with a spin of 1 carry the forces between these matter particles, and the spin-0 Higgs boson provides them with mass. Particles with half-integer spins follow a mathematical model called Fermi-Dirac statistics and are known as fermions. Those with integer or zero spin follow Bose-Einstein statistics and are called bosons.

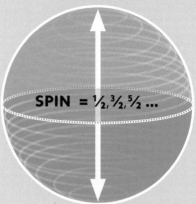

FERMIONS

SPIN = $\frac{1}{2}, \frac{3}{2}, \frac{5}{2}$ ...

Matter particles
The + or - sign on a particle's spin simply indicates whether its direction is up or down. When fermions join together their spins add up.

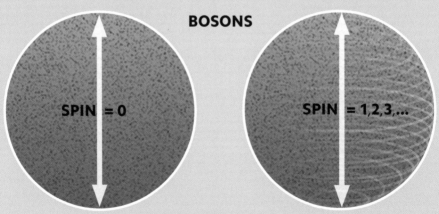

BOSONS

SPIN = 0

SPIN = 1, 2, 3, ...

Integer spin particles
Force-carrying bosons have a spin of 1, but larger bosons can be made by adding together fermions. The unique Higgs boson has zero spin (see p.127).

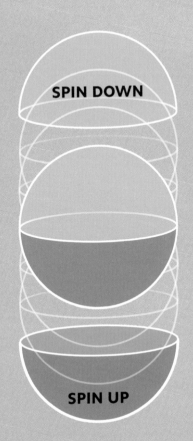

SPIN DOWN

SPIN UP

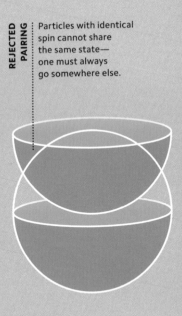

REJECTED PAIRING Particles with identical spin cannot share the same state— one must always go somewhere else.

Particle pairs
Two otherwise identical particles can have the same energy, momentum, and position if their spins are complementary.

NO ROOM FOR TWO

One of the most important rules underlying the structure of matter, the Pauli exclusion principle prevents fermions from falling into completely identical quantum states. This generates a form of pressure that keeps particles apart even when other repulsive forces such as electromagnetism fail (for instance, inside superdense collapsed stars known as white dwarfs and neutron stars). It also explains why only two electrons (with up and down spins) can occupy the same orbital subshell within an atom (see p.31).

In terms of classical physics, the forces
binding an atomic nucleus create an
energy barrier or "potential well."
Classically, a particle could only escape
a well if it had enough kinetic energy to
overcome the potential energy deficit.
Energy below this value would mean that
the particle was trapped forever.

WHAT BARRIER?

An effect known as tunneling explains how subatomic
particles can sometimes cross apparently insurmountable
barriers. During radioactive alpha decay (see p.113), for
example, a cluster of protons and neutrons spontaneously
breaks free from a larger atomic nucleus by overcoming the
binding energy that holds the nucleus together. Such a leap
is impossible in classical physics, but the edges of a quantum
wave function can reach beyond the barrier, allowing for a
small chance of the particle being found there.

A possible outcome

Over time there is a finite probability that a radioactive particle will decay, but the decay event itself is impossible to predict. Tunneling may provide a solution as to why particles sometimes, but not always, escape from an atomic nucleus.

TUNNELING THROUGH

The amplitude of a wave function decreases exponentially across a potential barrier, but it can still end up with a nonzero value on the other side of the barrier, and therefore there is a nonzero chance of the particle being found on the other side of the barrier.

BARRIER

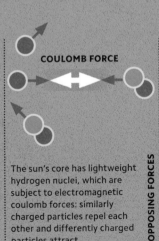

NUCLEI IN THE SUN

COULOMB FORCE

The sun's core has lightweight hydrogen nuclei, which are subject to electromagnetic coulomb forces: similarly charged particles repel each other and differently charged particles attract.

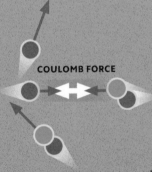

OPPOSING FORCES

COULOMB FORCE

Temperatures of millions of degrees give nuclei huge kinetic energy, but conditions inside the sun are not hot enough to account for the vast number of particles that overcome repulsion.

NUCLEAR FUSION

There are many more fusion reactions in the sun than would be expected given the coulomb barrier in place, and so quantum tunneling must play a key role.

DISTANCE NO OBJECT

One of the strangest quantum effects, entanglement arises from the nature of the quantum wave function itself. Pairs of particles produced in a single interaction share a common origin and cannot be described independently of one another. The particles can be separated by a vast distance, yet when the properties of one particle are measured, its entangled partner somehow instantaneously "knows," and itself collapses to the appropriate state.

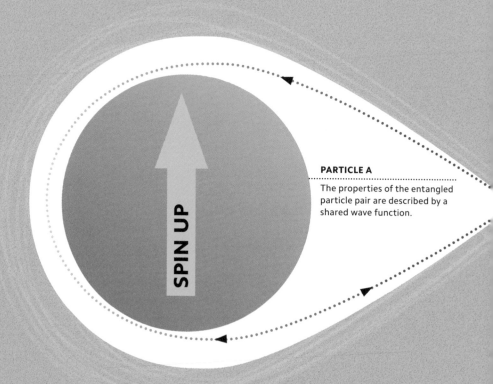

SPIN UP

PARTICLE A

The properties of the entangled particle pair are described by a shared wave function.

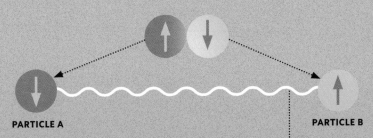

PARTICLE A

PARTICLE B

Correlation

A measurement of physical properties—such as spin, position, momentum, or polarization—of one particle in an entangled pair will correlate with the measurement in the second.

QUANTUM CONNECTION

Particles can be widely separated, yet measuring one resolves the properties of the other instantaneously.

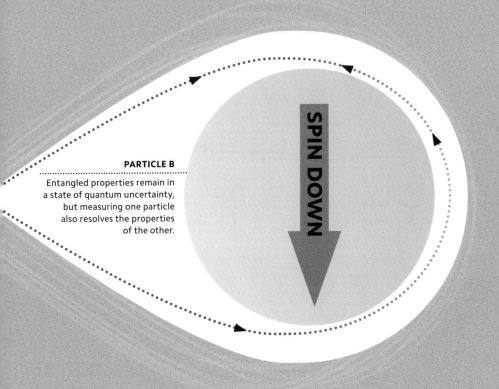

PARTICLE B

Entangled properties remain in a state of quantum uncertainty, but measuring one particle also resolves the properties of the other.

SPIN DOWN

QUANTUM TELEPORTATION

The strange physics of entanglement (see pp.72–73) can be put to work in quantum teleportation—the reconstruction of quantum information from one system in another system. Teleportation involves the creation of an entangled pair of quantum states (qubits; see p.106), which are then separated. The system to be teleported interacts with one qubit, resulting in a "classical" measurement that is then transmitted to the location of the other qubit at the speed of light, and used to reconstruct the interacting system.

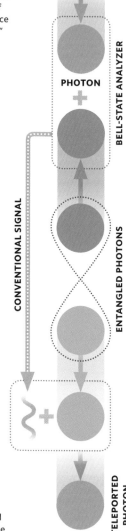

SENDER

Alice encodes information into a photon, which is then made to interact with one of a pair of entangled particles in a device called a "Bell-state analyzer."

PHOTON

BELL-STATE ANALYZER

CONVENTIONAL SIGNAL

ENTANGLED PHOTONS

Bob has the other half of the entangled photon pair. Information about Alice's system, sent by conventional means, allows him to recreate Alice's encoded photon and read its information.

RECEIVER

TELEPORTED PHOTON

COHERENCE

A quantum particle in complete
isolation can retain an indeterminate
state indefinitely.

ENVIRONMENTAL IMPACT

In a complex system, such as an atom,
interference with wave functions from
other particles causes indeterminate
systems to lose coherence.

ATOM

UNSTABLE ENVIRONMENT

Quantum systems in the indeterminate state
described by a wave function are said to be
"coherent." The phenomenon of decoherence is a loss
of quantum information as a system interacts with its
surroundings. Unless a system is perfectly isolated, its
coherence deteriorates through interactions with the
wave functions of its environment. Decoherence is a
major challenge for quantum computing systems
that rely on keeping particles in a state of
long-term coherence.

INSIDE SOLID OBJECTS

Individual atoms consist of positively charged nuclei surrounded by negatively charged electrons (see pp.10–11). In order to form large-scale solid materials, such as crystals, however, atoms surrender some of their electrons, allowing them to float freely (to a certain degree) within a fixed geometric lattice of positively charged ions. Quantum effects govern where electrons can reside—modeling the electrons as a "gas" of fermions provides insights into phenomena such as electrical and heat conduction and insulation.

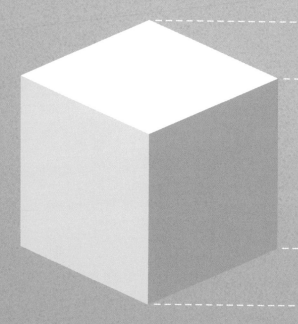

SOLID OBJECT

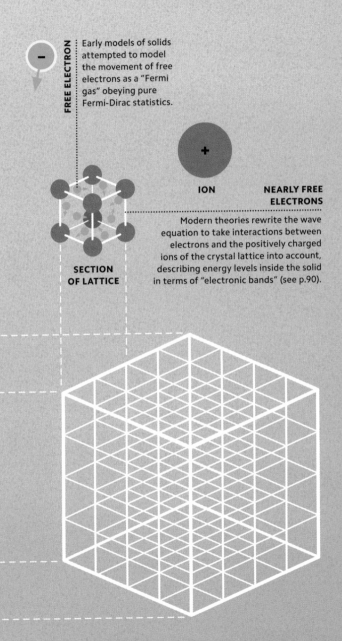

Early models of solids attempted to model the movement of free electrons as a "Fermi gas" obeying pure Fermi-Dirac statistics.

ION

SECTION OF LATTICE

NEARLY FREE ELECTRONS

Modern theories rewrite the wave equation to take interactions between electrons and the positively charged ions of the crystal lattice into account, describing energy levels inside the solid in terms of "electronic bands" (see p.90).

CRYSTALLINE STRUCTURE

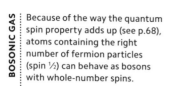

Because of the way the quantum spin property adds up (see p.68), atoms containing the right number of fermion particles (spin ½) can behave as bosons with whole-number spins.

As a sparse gas of bosonic atoms is cooled to very low temperatures, their wave functions expand and begin to overlap with each other.

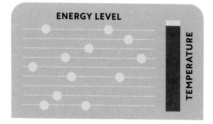

Normal temperatures
At high temperatures, the boson gas seems outwardly normal, with particles at a wide range of energy levels— although some levels are shared.

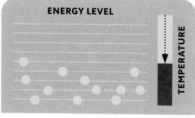

Extreme cooling
As temperatures fall and the particles lose kinetic energy, their range of possible energy states is reduced. More particles start to share the same state.

ALTERED STATES

When a large collection of bosons is cooled to a very low temperature, the range of possible energy states available to the particles is greatly diminished, producing a strange form of matter called a Bose-Einstein condensate (BEC). Because the Pauli exclusion principle (see p.69) does not apply to bosons, the particles are not forced to maintain separate states—instead they fall into a shared low-energy state described by a single wave equation.

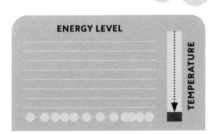

ENERGY LEVEL

TEMPERATURE

Critical temperature
Below a certain critical temperature, all the particles fall into the lowest possible energy state, described by a single quantum wave function.

BOSE-EINSTEIN CONDENSATE
As the temperature is reduced toward absolute zero (−459.67°F/ −273.15°C), the wave functions expand still further and merge to form a BEC that behaves as a single giant particle.

FLOWING WITHOUT FRICTION

Superfluidity is a strange phenomenon in which a liquefied gas cooled to an extremely low temperature flows without internal friction between its atoms. Superfluids arise when atoms fall into states where they are governed by Bose-Einstein statistics (see pp.78–79). Their frictionless flow allows them to "creep" up container walls, form curious wave patterns, and even to slow the speed of light.

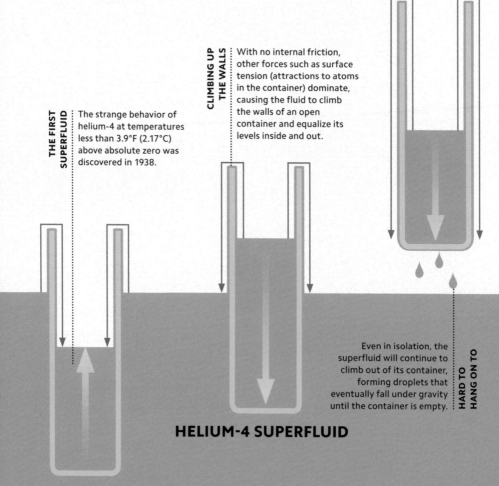

THE FIRST SUPERFLUID

The strange behavior of helium-4 at temperatures less than 3.9°F (2.17°C) above absolute zero was discovered in 1938.

CLIMBING UP THE WALLS

With no internal friction, other forces such as surface tension (attractions to atoms in the container) dominate, causing the fluid to climb the walls of an open container and equalize its levels inside and out.

HARD TO HANG ON TO

Even in isolation, the superfluid will continue to climb out of its container, forming droplets that eventually fall under gravity until the container is empty.

HELIUM-4 SUPERFLUID

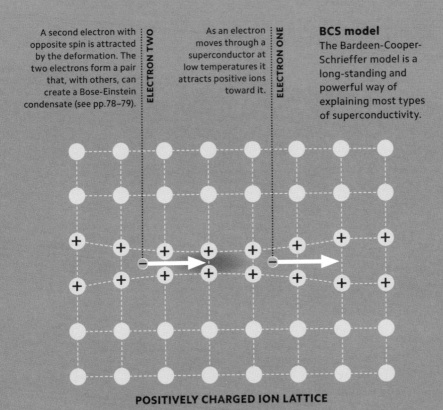

A second electron with opposite spin is attracted by the deformation. The two electrons form a pair that, with others, can create a Bose-Einstein condensate (see pp.78–79).

ELECTRON TWO

As an electron moves through a superconductor at low temperatures it attracts positive ions toward it.

ELECTRON ONE

BCS model
The Bardeen-Cooper-Schrieffer model is a long-standing and powerful way of explaining most types of superconductivity.

POSITIVELY CHARGED ION LATTICE

ENDLESS CHARGE

In normal electrical conductors, energy is lost to resistance—interactions between the electron particles that transmit current and their surroundings. However, at very low temperatures quantum behavior leads to an effect called superconductivity, in which current flows without resistance through certain materials. Electrons in a superconductor travel in Cooper pairs that exhibit the properties of bosons (see pp.78–79) rather than fermion particles, allowing them to flow without friction in a similar way to superfluids.

STRANGE QUANTUM ATOMS

Most atoms are in a state of constant motion, which makes their properties hard to measure precisely. To get the most accurate measurements, physicists cool them to extremely low temperatures, where the random movement of particles due to excess kinetic (motion) energy is minimized. This can be done by using laser beams to slow down the atoms, ultimately reaching temperatures close to absolute zero (−459.67°F/−273.15°C), where strange quantum properties can develop.

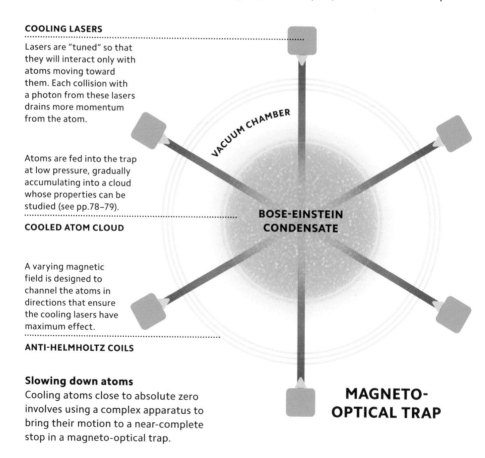

COOLING LASERS

Lasers are "tuned" so that they will interact only with atoms moving toward them. Each collision with a photon from these lasers drains more momentum from the atom.

Atoms are fed into the trap at low pressure, gradually accumulating into a cloud whose properties can be studied (see pp.78–79).

COOLED ATOM CLOUD

A varying magnetic field is designed to channel the atoms in directions that ensure the cooling lasers have maximum effect.

ANTI-HELMHOLTZ COILS

VACUUM CHAMBER

BOSE-EINSTEIN CONDENSATE

MAGNETO-OPTICAL TRAP

Slowing down atoms
Cooling atoms close to absolute zero involves using a complex apparatus to bring their motion to a near-complete stop in a magneto-optical trap.

RYDBERG ELECTRON

The wave function of the outer electron has very little overlap with those of the inner electrons and so it is fairly immune to interference from within.

NUCLEUS

Excited atom
The inner electrons of a Rydberg atom shield the outermost electrons from most of the influence of the nucleus, from their distant position the electric forces they experience are similar to those in a hydrogen atom.

RYDBERG ATOM

UNUSUAL ORBITS

Rydberg atoms are a form of matter in which at least one of an atom's electrons is boosted into a large orbit with a very high "principal quantum number." The great distance between the electron and the atom's inner core produces curious effects. Some properties of Rydberg atoms mirror those of hydrogen, but the atom's loose grip on the outer electron means it is easily ionized, and highly sensitive to electric and magnetic fields.

QUANT
TECHN

U M
O L O G Y

Quantum physics is at the heart of many established and emerging technologies. Applications that harness quantum principles in their design—ranging from everyday electronics to satellite timekeepers—can be broadly described as quantum technologies. The devices that laid the foundation of the Information Age are built on quantum physics—for instance, semiconductor devices are designed around the quantization of electron energy levels, which restrict the movement of charge in a lattice. In the near future, similarly transformative new technologies could be built on other quantum phenomena, such as entanglement.

FIRING PHOTONS AT ATOMS

When an electron falls from an excited state to its ground state (see p.30), it releases the energy difference as a photon. Stimulated emission involves firing photons at atoms, causing excited electrons to fall to their ground states and release new photons with wavelength and phase identical to the incident photons. These stimulate other atoms, resulting in a cascade of coherent light (see opposite).

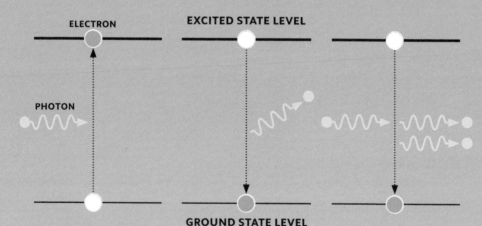

Absorption of light
When an electron absorbs energy from an incident photon, it may be excited (also referred to as being "pumped") to a higher energy state. An electron in an excited state may then decay to a lower energy state.

Spontaneous emission of light
When an electron falls to a lower state spontaneously, it releases energy in the form of a photon. The phase and direction of photons spontaneously emitted by electrons in a material is random.

Stimulated emission of light
An incident photon can cause an electron to fall to a lower energy state. In the process, the electron emits an additional photon, which has the same phase and direction as the incident photon.

HIGHLY CONCENTRATED

A laser is a device that emits light through the process of optical amplification, based on stimulated emission (see opposite). Unlike light from other sources, laser light is coherent, meaning that the waves are perfectly in step with each other and have the same frequency. The invention of lasers has allowed unprecedented control over light, with applications such as remote-sensing lidar, laser cutting, and spectroscopy. Lasers can also be used to trap and cool small particles, such as atoms and ions.

HIGH-INTENSITY FLASH LAMP
The ruby medium is pumped using a flashtube. Energy emitted from the flashtube is absorbed by electrons in the medium, which then jump to excited states.

ALL WAVES IN STEP
The laser emits rapid pulses of visible red light. Due to the photons being produced via stimulated emission, this light is perfectly coherent.

RUBY LASER

KEEPING TIME

Atomic clocks, which are used in technologies such as satellite navigation, keep precise time using the properties of certain atoms such as those of cesium-133. When atoms are exposed to photons, some electrons jump between energy levels. When the incident photons have precisely the same frequency as a cesium-133 atom, electrons in the atoms resonate and leap between energy levels. One second is defined as 9,192,631,770 oscillations at that frequency.

Measuring time

The most precise way of measuring time is based on using the frequency of microwave radiation that excites electrons to jump between energy states in certain atoms.

QUANTUM LEAP

The oscillator fires microwaves set to a specific frequency at the atoms, causing them to jump to a higher energy state.

CESIUM-133 ATOMS FIRED

Cesium atoms are ionized and fired through a magnetic gate that filters out any with a high-energy state. The low-energy-state atoms then continue on to the radio wave oscillator.

MAGNET

MAGNET

HIGH-ENERGY-STATE ATOMS REMOVED

The most accurate
atomic clocks will
not lose or gain more
than 1 second in
15 billion years.

FREQUENCY AND TIME
The definition of a second has been based on this frequency since 1968.

FEEDBACK TO OSCILLATOR
A second magnet filters out low-energy cesium atoms before a detector counts the number of atoms. If the detector counts enough high-energy-state atoms, then the oscillator is at the right frequency. If the number of high-energy atoms is too low, then the oscillator needs adjusting to the correct frequency.

9,192,631,770 OSCILLATIONS

RADIO WAVE SIGNAL
SENT AT 9,192,631,770 HZ

OSCILLATOR

MAGNET

DETECTOR

MAGNET

LOW-ENERGY-STATE
ATOMS REMOVED

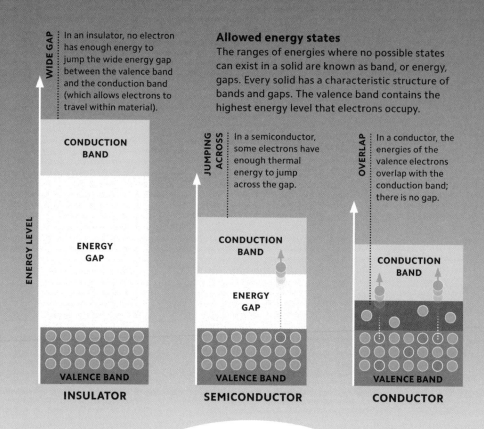

Allowed energy states
The ranges of energies where no possible states can exist in a solid are known as band, or energy, gaps. Every solid has a characteristic structure of bands and gaps. The valence band contains the highest energy level that electrons occupy.

WIDE GAP — In an insulator, no electron has enough energy to jump the wide energy gap between the valence band and the conduction band (which allows electrons to travel within material).

JUMPING ACROSS — In a semiconductor, some electrons have enough thermal energy to jump across the gap.

OVERLAP — In a conductor, the energies of the valence electrons overlap with the conduction band; there is no gap.

ENERGY LEVEL

CONDUCTION BAND

ENERGY GAP

VALENCE BAND

INSULATOR

CONDUCTION BAND

ENERGY GAP

VALENCE BAND

SEMICONDUCTOR

CONDUCTION BAND

VALENCE BAND

CONDUCTOR

SOLID STATES

In order to understand the electric properties of materials, it is necessary to understand how electrons move within them. Band theory is a model that describes the permitted and forbidden energy levels—bands and band gaps—in materials with solid structures (see pp.76–77), which restrict the behavior of electrons. This model explains thermal and electric properties of solid matter and is the basis of solid-state electronic devices, such as transistors (see opposite), diodes (see p.92), and solid-state data storage devices.

SILICON CHIPS

A transistor is a device that changes electronic signals. It is made from silicon, which is "doped" to give it different properties. For example, when electrons are removed, "holes" are left for electrons to flow into. Putting together layers of silicon, each with either an excess (n-type) or deficit (p-type) of electrons, creates transistors that can amplify or switch a current. Many transistors are integrated onto a single computer chip.

Basic transistor

In an n-p-n transistor, a layer with a deficit of electrons (p-type) is sandwiched between layers with excess electrons (n-type). Excess electrons can flow into the p-type region.

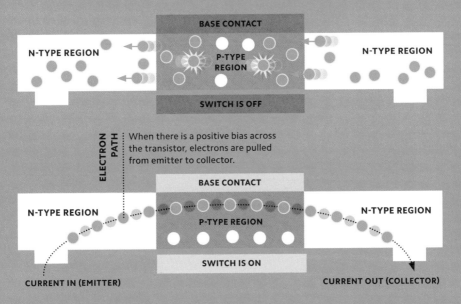

BASE CONTACT

N-TYPE REGION

P-TYPE REGION

N-TYPE REGION

SWITCH IS OFF

ELECTRON PATH | When there is a positive bias across the transistor, electrons are pulled from emitter to collector.

BASE CONTACT

N-TYPE REGION

P-TYPE REGION

N-TYPE REGION

SWITCH IS ON

CURRENT IN (EMITTER)

CURRENT OUT (COLLECTOR)

LIGHT RELEASE

A light-emitting diode (LED) emits light when a current is applied. In an LED, a p-type and an n-type semiconductor are placed close together, between electrical contacts. When current flows, excess electrons flow from the n-type to the p-type layer, where they fall into "holes" in the semiconductor and release photons (light). The color of the light emitted depends on the energy required for electrons to cross the band gap (see p.90).

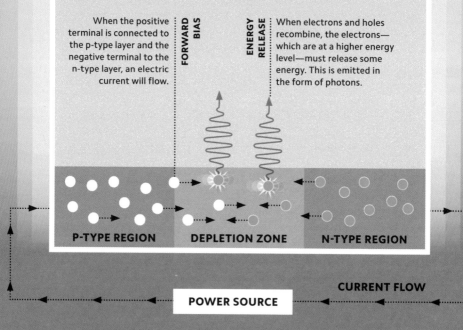

When the positive terminal is connected to the p-type layer and the negative terminal to the n-type layer, an electric current will flow.

FORWARD BIAS

ENERGY RELEASE

When electrons and holes recombine, the electrons—which are at a higher energy level—must release some energy. This is emitted in the form of photons.

P-TYPE REGION   **DEPLETION ZONE**   **N-TYPE REGION**

CURRENT FLOW

POWER SOURCE

CAPTURING PHOTONS

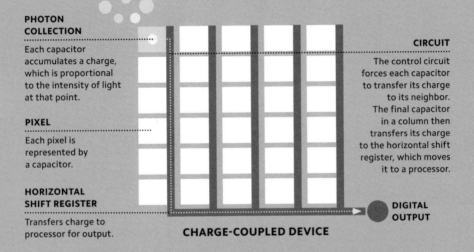

PHOTON COLLECTION

Each capacitor accumulates a charge, which is proportional to the intensity of light at that point.

PIXEL

Each pixel is represented by a capacitor.

HORIZONTAL SHIFT REGISTER

Transfers charge to processor for output.

CIRCUIT

The control circuit forces each capacitor to transfer its charge to its neighbor. The final capacitor in a column then transfers its charge to the horizontal shift register, which moves it to a processor.

DIGITAL OUTPUT

CHARGE-COUPLED DEVICE

A charge-coupled device (CCD) is a circuit of linked capacitors, which store and transfer electrical charge. CCD image sensors convert light into electrical signals and are important components in digital imaging. Pixels are represented by semiconductor capacitors that free electrons when photons are absorbed, accumulating electrical charge proportional to light intensity at that point. A control circuit transfers charges through a capacitor array, converting them into a readable signal.

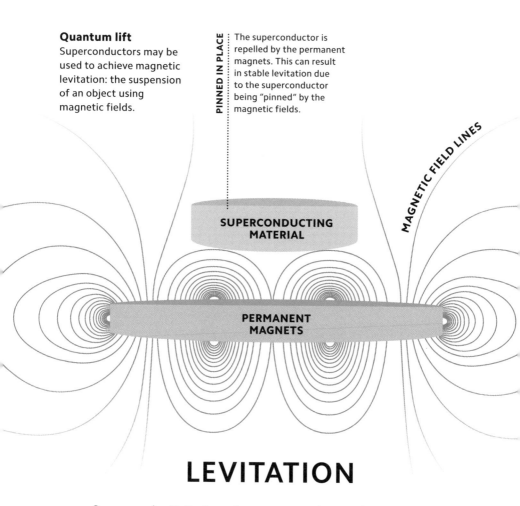

Quantum lift
Superconductors may be used to achieve magnetic levitation: the suspension of an object using magnetic fields.

PINNED IN PLACE

The superconductor is repelled by the permanent magnets. This can result in stable levitation due to the superconductor being "pinned" by the magnetic fields.

MAGNETIC FIELD LINES

SUPERCONDUCTING MATERIAL

PERMANENT MAGNETS

LEVITATION

Superconductivity is a phenomenon observed near absolute zero (−459.67°F/−273.15°C), in which electrical resistance vanishes (see p.81). Superconducting electromagnets are made from superconducting wire coils that conduct massive currents, generating powerful magnetic fields. When a superconductor is in a magnetic field, small currents are created. These currents then create opposite magnetic fields, expelling the field and preventing the magnetic field from penetrating the superconductor.

TUNNELING PAIRS

The Josephson effect is a rare example of a quantum phenomenon observable on a macroscopic scale. It involves an electric current flowing indefinitely, without an applied voltage, across a Josephson junction. A Josephson junction is a pair of superconductors with a thin barrier sandwiched between them. Electrons in Cooper pairs (see p.81) tunnel through the barrier from one superconductor to the other with no resistance. Josephson junctions are used in SQUIDS (see pp.96–97).

COOPER PAIR

Across the barrier
Pairs of electrons tunnel through the barrier from one superconductor to the other until a critical current is reached and a voltage appears across the junction.

BARRIER

SUPERCONDUCTOR

THIN BARRIER

A very thin barrier of insulating or normal conduction material between two superconductors forces the pair to tunnel (see pp.70–71) across.

COOPER PAIR

SQUIDS

A superconducting quantum interference device (SQUID) is used to detect extremely faint magnetic fields, including signals associated with neural activity. A SQUID is based on a superconducting loop containing two Josephson junctions (see p.95). In a magnetic field, the current in the ring varies to shift the magnetic flux (number of field lines) within it to an energetically preferable value, causing voltage to vary with the magnetic field.

SEMICONDUCTOR

CURRENT

In the presence of a small magnetic field, a current is created (induced). The current generates a magnetic field. If the field outside the loop changes, the current in the loop increases or decreases the magnetic flux inside the loop to an energetically preferable value (a multiple of the flux quantum). This produces a measurable change in voltage across the junctions.

INDUCED CURRENT

JOSEPHSON JUNCTION

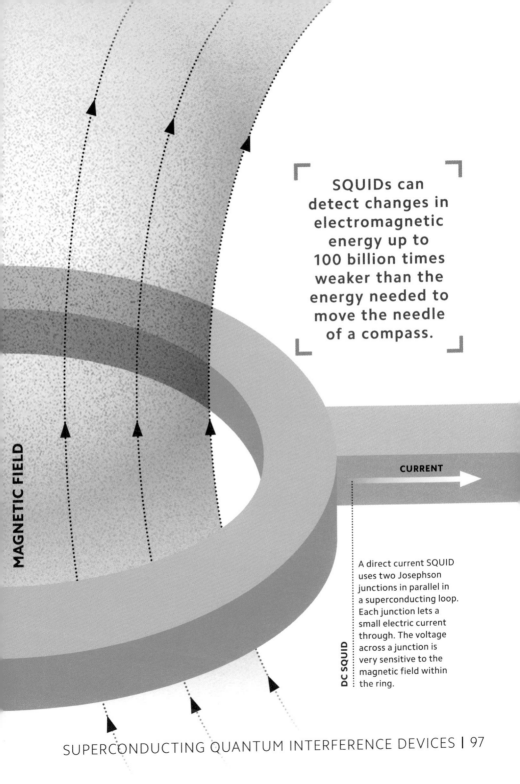

SQUIDs can detect changes in electromagnetic energy up to 100 billion times weaker than the energy needed to move the needle of a compass.

MAGNETIC FIELD

CURRENT

DC SQUID

A direct current SQUID uses two Josephson junctions in parallel in a superconducting loop. Each junction lets a small electric current through. The voltage across a junction is very sensitive to the magnetic field within the ring.

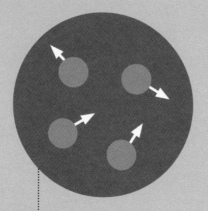

RANDOM POSITION

With no external magnetic field, protons spin inside the body with the axes of their own magnetic fields randomly aligned.

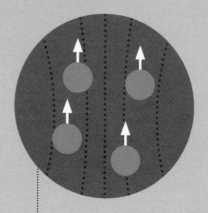

MOSTLY ALIGNED

When a powerful external magnetic field is applied, the axis of each proton's magnetic field lines up (parallel) with this magnetic field.

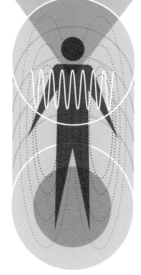

MRI SCANNER

Quantum scanner

When the spin of a single proton is measured, it can be in one of two states: parallel or antiparallel. During an MRI scan, protons switch from parallel to antiparallel and back again.

> "It was eerie. I saw myself in that machine. I never thought my work would come to this."
>
> Isidor Isaac Rabi,
> whose work led to the MRI

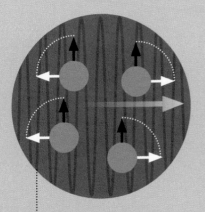

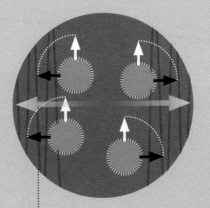

Radio wave pulses of a specific frequency are sent through the body, causing the protons to vibrate and exciting the protons into a different alignment (antiparallel).

When the radio pulses stop, the protons keep vibrating and gradually realign with the magnetic field. Their vibration releases measurable radio waves (energy) in the process.

LOOKING INSIDE

Magnetic resonance imaging (MRI) is a noninvasive medical imaging technique. The patient is placed in a powerful magnetic field, forcing the magnetic moments (see p.67) of protons in the body to align. Protons make up the nuclei of atoms of hydrogen, one of the most abundant elements in the body. When a current is applied, the protons become excited and spin (or vibrate) out of position. Measuring the frequency of the vibrations of the protons after the current is switched off allows different tissues to be distinguished.

SEEING WITHOUT LIGHT

The resolution of a microscope is limited by wavelength. When it comes to observing tiny objects, matter waves (which have shorter wavelengths) can be more helpful than light. The wavelength of an electron is up to 100,000 times shorter than that of visible light, allowing for the imaging of up to 100,000 times smaller objects, such as viruses. Scanning electron microscopes use beams of accelerated electrons to illuminate objects, bent by magnetic fields rather than glass lenses.

SECONDARY DETECTOR
A secondary electron detector attracts and registers electrons emitted from the sample.

Beaming electrons
A focused electron beam scatters from atoms on an object's surface, producing signals that allow for a very detailed image to be built.

REFLECTED BEAMS
Some high-energy electrons are reflected or backscattered from atoms within the specimen.

ELECTRON BEAM

MAGNETS

ELECTRON DETECTOR

ELECTRON DETECTION

A detector records the intensity of scattered electrons in each position to build an image of the specimen.

GUIDING MAGNET

A magnetic field bends the beam of electrons, much like the way a glass lens bends a beam of light.

"In basic research, the use of the electron microscope has revealed to us the complex universe of the cell, the basic unit of life."

Günter Blobel

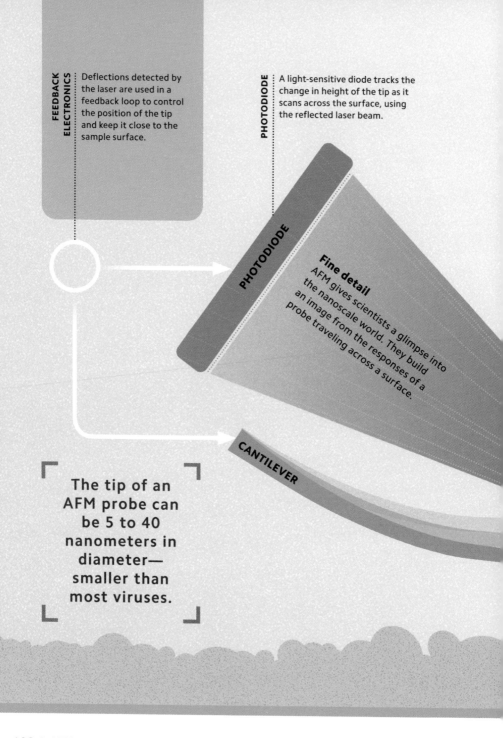

Deflections detected by the laser are used in a feedback loop to control the position of the tip and keep it close to the sample surface.

PHOTODIODE

A light-sensitive diode tracks the change in height of the tip as it scans across the surface, using the reflected laser beam.

PHOTODIODE

Fine detail

AFM gives scientists a glimpse into the nanoscale world. They build an image from the responses of a probe traveling across a surface.

CANTILEVER

The tip of an AFM probe can be 5 to 40 nanometers in diameter—smaller than most viruses.

ATOMIC PROBE

Atomic force microscopy (AFM) is a scanning technique that involves "touching" the surfaces of objects. When a cantilever with an extremely sharp probe approaches the sample's surface, forces between the surface and the probe cause the cantilever to be deflected slightly. This movement is detected with a laser. AFM allows for nanometer-scale resolution, and can be used for imaging and manipulating individual atoms.

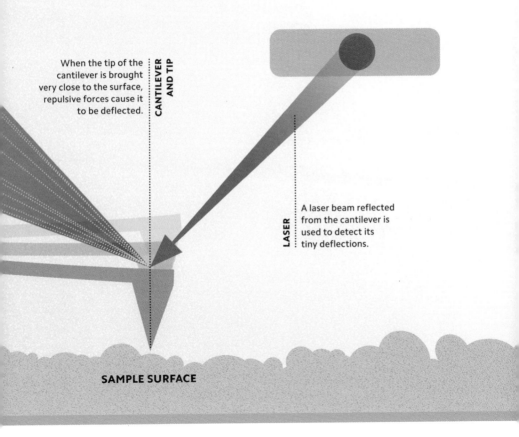

When the tip of the cantilever is brought very close to the surface, repulsive forces cause it to be deflected.

CANTILEVER AND TIP

A laser beam reflected from the cantilever is used to detect its tiny deflections.

LASER

SAMPLE SURFACE

QUANTU
INFORM

M
ATION

In 1980, well before computers became household objects, the American physicist Paul Benioff demonstrated that a computer could theoretically operate under the laws of quantum physics. Today, there are many models of quantum computing. Quantum computers store and manipulate quantum information to perform calculations, potentially carrying out new calculations and algorithms that would offer an exponential time improvement on today's classical computers. Although serious technical barriers stand in the way of quantum computers entering the mainstream, they are expected to transform computing, communication, and security in the 21st century.

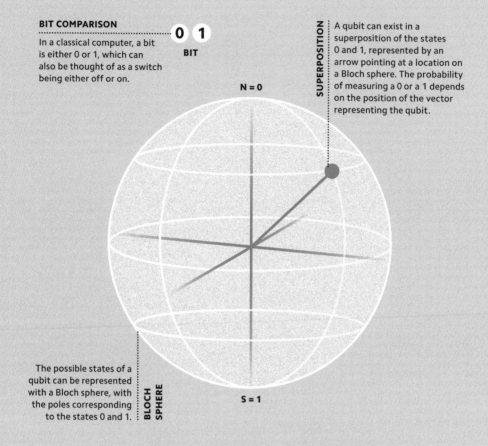

BIT COMPARISON

In a classical computer, a bit is either 0 or 1, which can also be thought of as a switch being either off or on.

0 1

BIT

N = 0

SUPERPOSITION

A qubit can exist in a superposition of the states 0 and 1, represented by an arrow pointing at a location on a Bloch sphere. The probability of measuring a 0 or a 1 depends on the position of the vector representing the qubit.

The possible states of a qubit can be represented with a Bloch sphere, with the poles corresponding to the states 0 and 1.

BLOCH SPHERE

S = 1

NOT JUST ZEROS AND ONES

The basic unit of information in classical computing is the bit, which can take one of two states (represented as 0 or 1). The quantum equivalent, known as a "qubit," is not in one state or the other, but instead in a superposition (see pp.38–39) of these two states. Qubits can be encoded in the properties of particles such as electron spin (up or down) and photon polarization (vertical or horizontal).

SUPERFAST

Quantum computers use quantum processes to store and manipulate data. They can perform certain calculations—such as breaking down very large numbers into their prime factors—much faster than classical computers. In theory, quantum computers have the ability to execute tasks that would be practically impossible on a classical computer. However, there are serious technical challenges associated with building quantum computers, most significantly in maintaining the wave function of the qubits (see p.75).

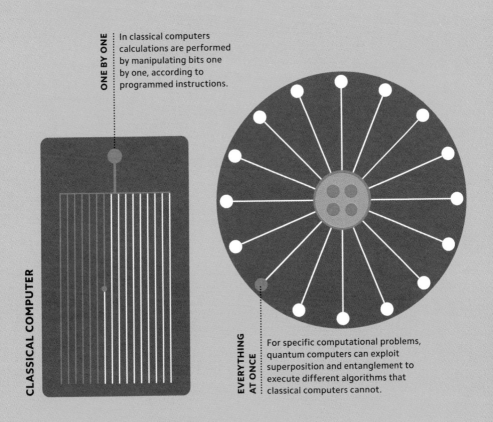

ONE BY ONE In classical computers calculations are performed by manipulating bits one by one, according to programmed instructions.

CLASSICAL COMPUTER

EVERYTHING AT ONCE For specific computational problems, quantum computers can exploit superposition and entanglement to execute different algorithms that classical computers cannot.

QUANTUM CODES

When data is encrypted, only a person with the key can unscramble and read it. Quantum cryptography uses quantum phenomena such as superposition (see pp.38–39) or entanglement (see pp.72–73) to encrypt and decrypt data. This can make it impossible to snoop on encrypted conversations; an eavesdropper must measure a quantum key to gain information about it, causing the wave functions of the shared quantum keys to collapse, thus revealing the unauthorized access.

Secret codes

Alice shares a secret message with Bob using a series of polarized photons. Alice and Bob then agree on a "sifted key," which is based on measurements made with compatible polarization filters.

Bob uses polarizing filters and a detector to obtain information from the photons.

Photons are polarized using filters.

To create the sifted key, Bob tells Alice the sequence of the filters he used and they discard the measurements that do not match.

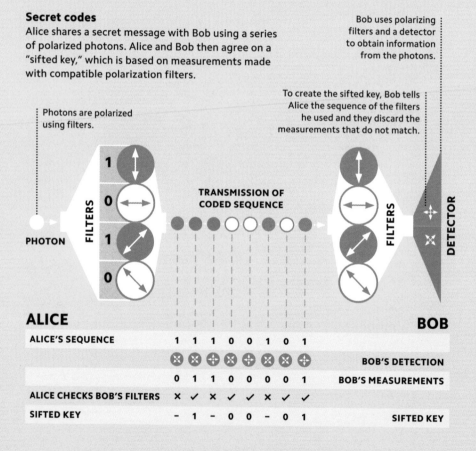

ALICE									BOB
ALICE'S SEQUENCE	1	1	1	0	0	1	0	1	
	⊗	⊗	⊕	⊗	⊗	⊕	⊗	⊕	**BOB'S DETECTION**
	0	1	1	0	0	0	0	1	**BOB'S MEASUREMENTS**
ALICE CHECKS BOB'S FILTERS	✗	✓	✗	✓	✓	✗	✓	✓	
SIFTED KEY	–	1	–	0	0	–	0	1	**SIFTED KEY**

SIMULATORS

Some systems are too complex for even supercomputers to simulate—particularly systems with properties that cannot be classically simulated, such as entanglement. However, these can be simulated with analog systems of real particles with quantum properties, such as ultracold gases. While quantum computers can theoretically one day be programmed to solve any problem, quantum simulators are already used to explore specific problems.

Ions are arranged in a matrix using magnetic fields and laser beams.

TOP VIEW IMAGE ION MATRIX

DETECTOR LENS

IONS SUSPENDED IN A MAGNETIC FIELD

Electric and magnetic fields trap hundreds of ions in an orderly 2D lattice.

COOLING LASER BEAM

LASER BEAM

LASERS
Highly sensitive laser beams are used for measuring the ions' properties, such as temperatures.

LASER BEAM

Trapped ion simulator
It may be possible to simulate interactions in quantum magnetism using trapped ions that mimic this magnetic behavior. This would not be feasible using a classical computer.

NUCLE
PHYSI

A R C S

On the eve of the 20th century, the discovery of radiation emerging from within atoms challenged the long-held belief that atoms—which had been named for their seeming indivisibility—were the fundamental building blocks of matter. This led to the discovery of atomic nuclei and marked the beginning of the exploration of the subatomic world. At this scale, quantum physics is required to explain how natural phenomena occur. The field of nuclear physics involves the study of nuclei (the dense, positively charged objects found at the center of atoms), their constituents, and associated phenomena such as radioactivity, fission, and fusion.

A neutron may also be ejected from the atomic nucleus during decay.

UNSTABLE NUCLEUS

A nucleus is radioactive if it spontaneously decays (emits radiation) to become a new, more stable nucleus.

Radioactivity was discovered by the French scientist Henri Becquerel in 1896 during experiments with X-rays and phosphorescent material.

β **BETA PARTICLE**

In beta minus decay, a neutron transforms into a proton, ejecting an electron and electron antineutrino from the nucleus. In beta plus decay, a proton transforms into a neutron, ejecting a positron and an electron neutrino in the process.

α ALPHA PARTICLE

During alpha decay, an alpha particle (a helium nucleus comprising two protons and two neutrons) is emitted from the unstable parent nucleus.

γ GAMMA RAYS

High-energy photons are emitted during gamma decay.

SEEKING STABILITY

Radioactivity is the emission of waves or particles from an unstable atomic nucleus. This occurs when a nucleus spontaneously transforms into a more stable configuration itself by emitting energy. The random nature of the quantum world means it is impossible to predict when an individual nucleus will decay, although the decay of a large group of identical nuclei can be described by their half-life (time taken for half of the nuclei to decay).

Nuclear reaction

Nuclear fission is initiated by bombarding radioactive material (most commonly uranium-235 in nuclear reactors) with neutrons.

NUCLEUS SPLITS

When a neutron destabilizes the uranium nucleus, the nucleus typically splits into two parts.

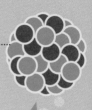

NEUTRON

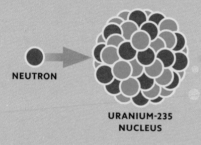

URANIUM-235 NUCLEUS

ENERGY RELEASED

SPLITTING ATOMS

Nuclear fission is the splitting of an atomic nucleus into smaller pieces. When added together, all of the lower-mass fragments left after fission weigh less than the original heavy nucleus—the missing mass is released as energy. Neutrons released in fission can go on to collide with other nuclei and can cause them to undergo fission. This is called a chain reaction (such as in a nuclear reactor or an atomic weapon).

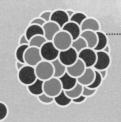

STARTING A CHAIN REACTION
When neutrons released in fission strike further uranium nuclei, this can lead to a chain of nuclear reactions.

NEUTRON

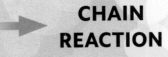

CHAIN REACTION

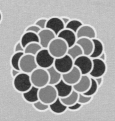

"The unleashed power of the atom has changed everything save our modes of thinking and we thus drift toward unparalleled catastrophe."
Albert Einstein

HELIUM

The two nuclei merge to produce a helium nucleus. A vast amount of energy is also released.

DEUTERIUM

Deuterium is a hydrogen nucleus with an extra neutron.

FUSION

NUCLEUS OF AN ATOM

TRITIUM

Tritium is a hydrogen isotope that has two extra neutrons in its nucleus.

NEUTRON EMISSION

An excess neutron is also emitted during fusion.

ENERGY

CREATING ELEMENTS

Stellar fusion is responsible for the variety of elements in the universe. Heavy elements are created as dying stars collapse, in the merging of neutron stars, or during other high-energy astrophysical events.

COMBINING NUCLEI

Two or more nuclei combine to form a larger nucleus in a process called nuclear fusion. When light nuclei fuse, they lose a little mass, which is released as energy. The repulsive Coulomb forces between positively charged nuclei make fusion impossible under all but the most extreme conditions, such as inside stars. Under these conditions, nuclei can tunnel (see p.71) through the Coulomb barrier and be brought close enough to fuse.

"I would like nuclear fusion to become a practical power source. It would provide an inexhaustible supply of energy, without pollution or global warming."
Stephen Hawking

PARTI

PHYSI

C L E
C S

Particle physics involves the study of the most fundamental objects and forces in nature. In the 20th century, the discovery of a "zoo" of elementary and composite particles allowed scientists to build the most successful theory yet for understanding particle physics: the Standard Model. This proposes that matter is made up of 12 fundamental particles (fermions), while the three quantum forces—strong, weak, and electromagnetic—are carried by force-carrying particles (bosons). According to quantum field theory, all these particles are excitations of their underlying quantum fields.

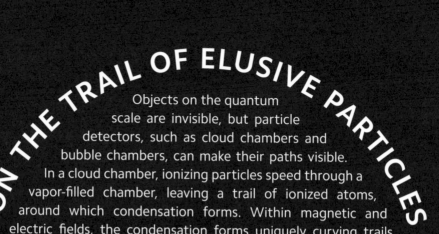

ON THE TRAIL OF ELUSIVE PARTICLES

Objects on the quantum scale are invisible, but particle detectors, such as cloud chambers and bubble chambers, can make their paths visible. In a cloud chamber, ionizing particles speed through a vapor-filled chamber, leaving a trail of ionized atoms, around which condensation forms. Within magnetic and electric fields, the condensation forms uniquely curving trails, which allow characteristics such as charge and mass to be calculated.

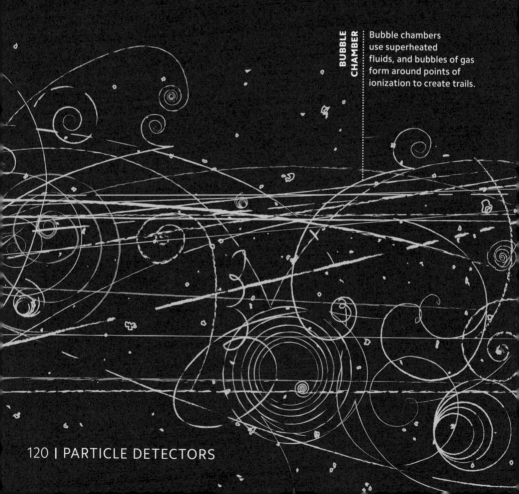

BUBBLE CHAMBER Bubble chambers use superheated fluids, and bubbles of gas form around points of ionization to create trails.

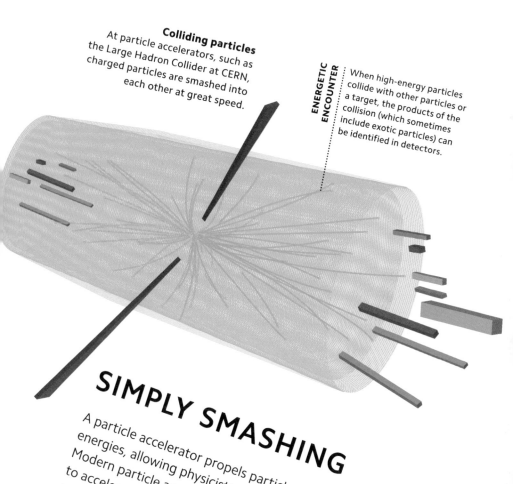

Colliding particles
At particle accelerators, such as the Large Hadron Collider at CERN, charged particles are smashed into each other at great speed.

ENERGETIC ENCOUNTER
When high-energy particles collide with other particles or a target, the products of the collision (which sometimes include exotic particles) can be identified in detectors.

SIMPLY SMASHING

A particle accelerator propels particles to extreme speeds and energies, allowing physicists to probe the smallest objects in nature. Modern particle accelerators use changing electromagnetic fields to accelerate and guide charged particles to almost light speed, and smash them into targets. Smashing particles together causes them to break apart, fuse, and—at extreme energies—create the exotic matter that briefly existed in the instant after the Big Bang.

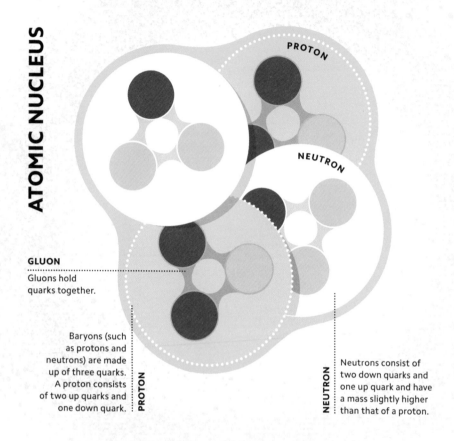

ATOMIC NUCLEUS

PROTON

NEUTRON

GLUON
Gluons hold
quarks together.

Baryons (such
as protons and
neutrons) are made
up of three quarks.
A proton consists
of two up quarks and
one down quark.

PROTON

NEUTRON

Neutrons consist of
two down quarks and
one up quark and have
a mass slightly higher
than that of a proton.

SMALLER THAN
AN ATOM

Atoms are built from elementary matter particles, such as quarks.
There are six quark flavors: up, down, strange, charm, top, and
bottom. Quarks also have a unique property known as "color
charge" (see p.135), which is unrelated to color in the everyday
sense. Particles with color charge cannot be isolated, so quarks
always combine to form "colorless" composite particles such as
protons, which are bound together through the strong force.

NO STRONG INTERACTIONS

Leptons are the other type of elementary matter particle. There are six flavors of leptons, divided into three generations and two classes: charged leptons (electron, muon, and tau), and electrically neutral neutrinos (electron neutrino, muon neutrino, and tau neutrino). All quarks and leptons have half-integer spin. Leptons are unaffected by the strong force. While charged leptons frequently interact with other particles throught the electromagnetic force, neutrinos are considered "ghostlike," barely interacting with anything as they can influence other particles only through the weak interaction.

NEUTRINOS

JUST PASSING THROUGH

About 100 trillion neutrinos pass through your body each second, arriving from the upper atmosphere, fusion processes in the sun, and from many other sources throughout the cosmos.

THE QUANTUM WORLD EXPLAINED

The Standard Model is the most successful theory for organizing the quantum world. It describes everything in terms of interactions played out by the set of elementary matter particles (fermions), force-carrying particles (gauge bosons), and the Higgs boson. Despite the success of the theory in predicting experimental results, it is considered a work in progress as there are some phenomena that cannot yet be explained (see pp.130–31).

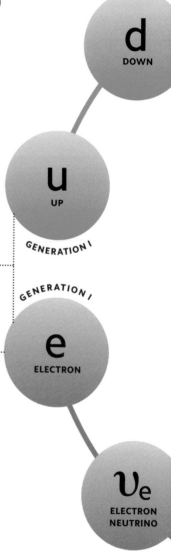

d
DOWN

u
UP

GENERATION I

GENERATION I

e
ELECTRON

υₑ
ELECTRON NEUTRINO

CREATING MATTER
....
Leptons and quarks are fermions (with half-integer spin). They are the smallest building blocks of matter.

LEPTONS
....
Leptons come in three "generations" and two types: charged and neutral. The charged leptons are the electron, muon, and tau. The neutrinos are neutral.

"The Standard Model is so complex it would be hard to put it on a T-shirt—though not impossible; you'd just have to write kind of small."
Steven Weinberg

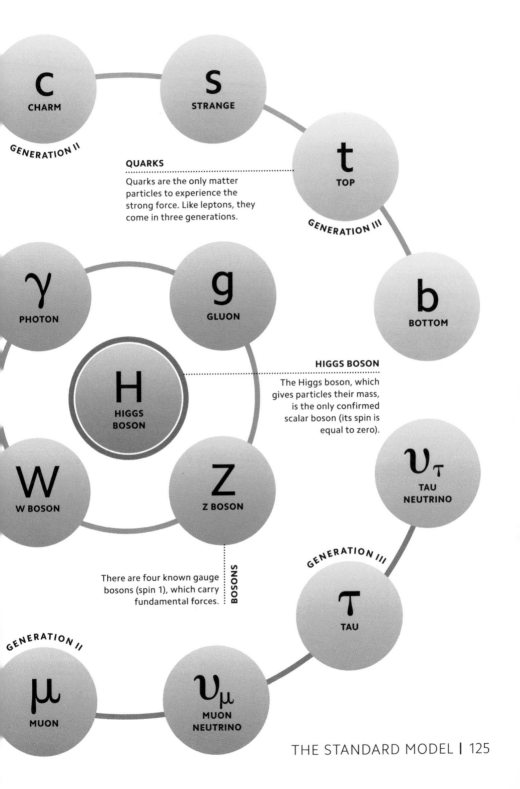

C CHARM

GENERATION II

S STRANGE

t TOP

GENERATION III

QUARKS

Quarks are the only matter particles to experience the strong force. Like leptons, they come in three generations.

γ PHOTON

g GLUON

b BOTTOM

H HIGGS BOSON

HIGGS BOSON

The Higgs boson, which gives particles their mass, is the only confirmed scalar boson (its spin is equal to zero).

W W BOSON

Z Z BOSON

υ_τ TAU NEUTRINO

There are four known gauge bosons (spin 1), which carry fundamental forces.

BOSONS

GENERATION III

τ TAU

GENERATION II

μ MUON

υ_μ MUON NEUTRINO

FORCE CARRIERS

Quantum physics successfully incorporates three of the four fundamental forces: electromagnetic, strong, and weak. When particles interact, they do so by exchanging the gauge (spin 1) bosons, also known as "force carriers," associated with those forces. The electromagnetic force is mediated by photons, the weak force by W and Z bosons, and the strong force by gluons.

SPIN 1 BOSONS

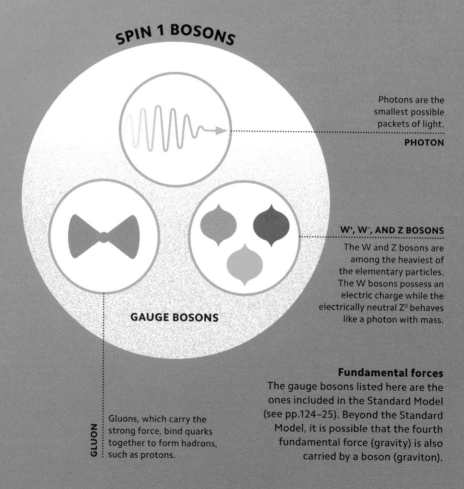

Photons are the smallest possible packets of light.

PHOTON

W⁺, W⁻, AND Z BOSONS

The W and Z bosons are among the heaviest of the elementary particles. The W bosons possess an electric charge while the electrically neutral Z⁰ behaves like a photon with mass.

GAUGE BOSONS

Fundamental forces
The gauge bosons listed here are the ones included in the Standard Model (see pp.124–25). Beyond the Standard Model, it is possible that the fourth fundamental force (gravity) is also carried by a boson (graviton).

GLUON
Gluons, which carry the strong force, bind quarks together to form hadrons, such as protons.

Higgs interactions

The interaction of particles with the Higgs field inhibits the particle's movement, generating their mass. Without it, all particles would zip around at the speed of light.

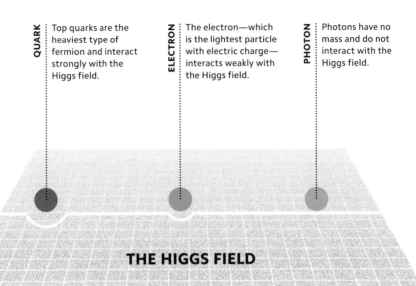

QUARK Top quarks are the heaviest type of fermion and interact strongly with the Higgs field.

ELECTRON The electron—which is the lightest particle with electric charge— interacts weakly with the Higgs field.

PHOTON Photons have no mass and do not interact with the Higgs field.

THE HIGGS FIELD

WHY PARTICLES HAVE MASS

The Standard Model includes one scalar (spin zero) boson, which is called the Higgs Boson. This is the particle associated with the Higgs field: a field that permeates throughout all space and gives particles their mass. The more strongly a particle interacts with the Higgs field, the more mass it has. A particle that does not interact with the field at all (such as a photon) has no mass.

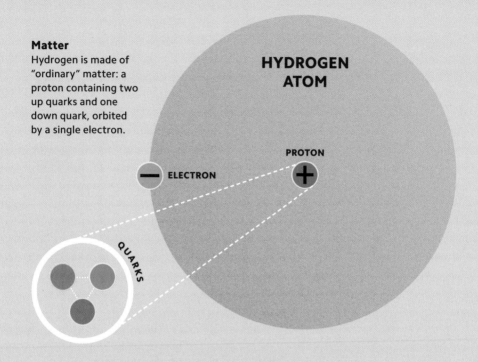

Matter
Hydrogen is made of
"ordinary" matter: a
proton containing two
up quarks and one
down quark, orbited
by a single electron.

HYDROGEN
ATOM

PROTON

ELECTRON

QUARKS

THE OPPOSITE
OF MATTER

Antimatter is composed of antiparticles, which have the same
mass but opposite electric charge (and some other properties)
to their corresponding "ordinary" matter particles. For example,
a positron has the same mass as an electron; however, it has
the same size, but opposite, charge to the electron. When a
particle meets its antimatter partner, they annihilate each
other with a burst of energy. The dominance of ordinary
matter in the universe is an unresolved mystery in physics.

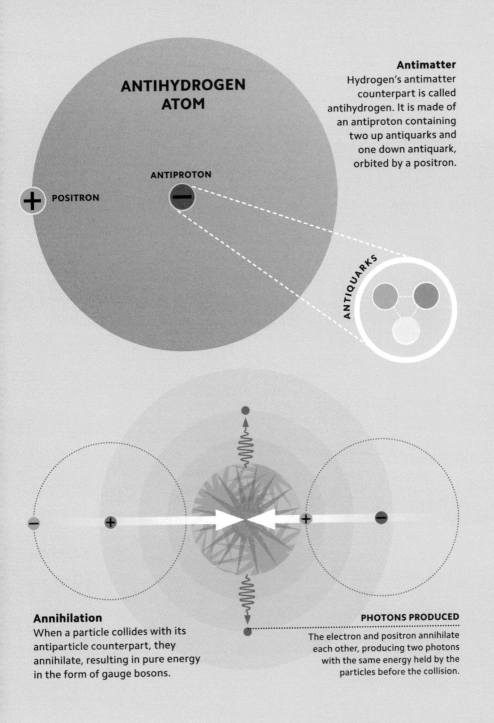

ANTIHYDROGEN ATOM

POSITRON

ANTIPROTON

ANTIQUARKS

Antimatter

Hydrogen's antimatter counterpart is called antihydrogen. It is made of an antiproton containing two up antiquarks and one down antiquark, orbited by a positron.

Annihilation

When a particle collides with its antiparticle counterpart, they annihilate, resulting in pure energy in the form of gauge bosons.

PHOTONS PRODUCED

The electron and positron annihilate each other, producing two photons with the same energy held by the particles before the collision.

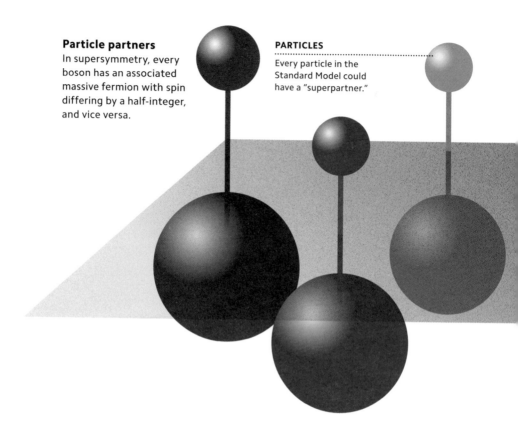

Particle partners
In supersymmetry, every boson has an associated massive fermion with spin differing by a half-integer, and vice versa.

PARTICLES
·······································
Every particle in the Standard Model could have a "superpartner."

NOT SO STANDARD

The Standard Model leaves many mysteries unresolved; for instance, it does not incorporate gravity or dark matter. Supersymmetry is a proposed extension that predicts a supersymmetric partner for every particle, with identical quantum numbers except spin, to resolve some problems with the model (such as by providing a dark matter candidate). Scientists explore beyond the Standard Model through high-energy particle accelerator experiments (see p.121).

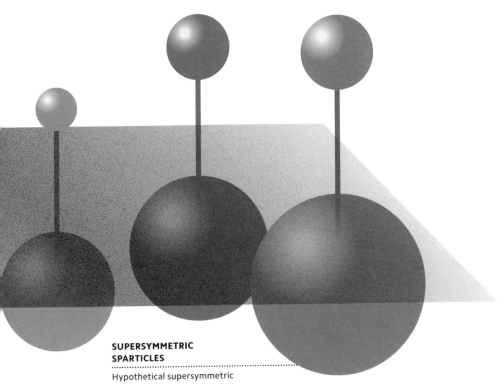

SUPERSYMMETRIC SPARTICLES

Hypothetical supersymmetric particles are named after their partners: for example, squarks are the supersymmetric partners of quarks.

> "Most gravity has no known origin. Is it some exotic particle? Nobody knows. Is dark energy responsible for expansion of the universe? Nobody knows."
>
> Neil deGrasse Tyson

UNIVERSAL FIELDS

Quantum field theory (QFT) is a broad framework that treats all particles as ripples in their underlying quantum fields. For instance, an electron emerges when the electron field is excited beyond a certain limit. Due to the uncertainty principle (see pp.42–43), these fields constantly froth with particles and antiparticles appearing from nothingness and vanishing an instant later. This theoretical framework incorporates theories such as the Standard Model (see pp.124–25).

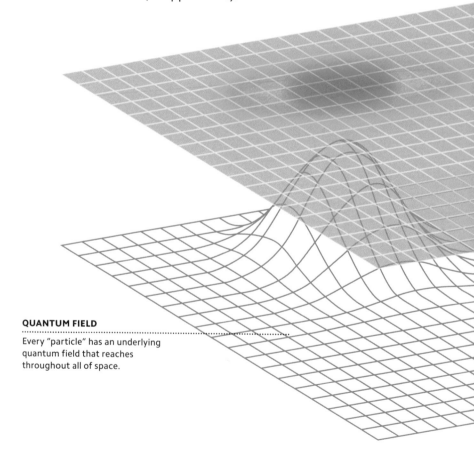

QUANTUM FIELD

Every "particle" has an underlying quantum field that reaches throughout all of space.

> "Quantum field theory, which was born ... from the marriage of quantum mechanics with relativity, is a beautiful but not very robust child."
>
> Steven Weinberg

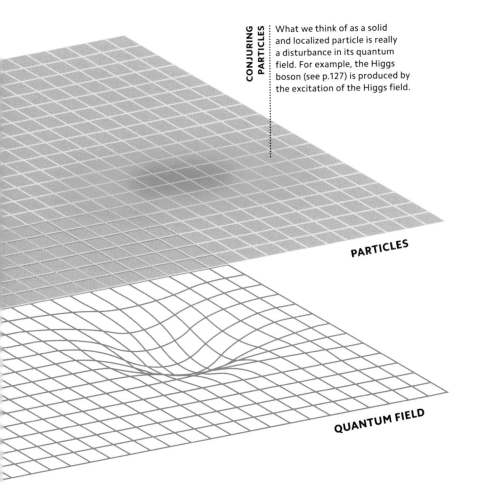

CONJURING PARTICLES What we think of as a solid and localized particle is really a disturbance in its quantum field. For example, the Higgs boson (see p.127) is produced by the excitation of the Higgs field.

PARTICLES

QUANTUM FIELD

THE JEWEL OF PHYSICS

ELECTRONS

Fermions, such as electrons and quarks, are represented with a straight, solid line.

INCOMING ELECTRON

OUTGOING ELECTRON

SINGLE PHOTON EXCHANGE

Photons and other gauge bosons are represented with wavy lines, except gluons, which are represented by curly lines (see opposite).

SPACE

Quantum electrodynamics (QED) is the quantum field theory for the electromagnetic force. It describes how electrically charged particles interact by exchanging photons: the force carrier for the electromagnetic force. As these photons are absorbed or released by charged particles, the energy exchanged by the photon changes the speed and direction of the particles. These processes can be visualized with Feynman diagrams.

VERTEX

Vertices represent interactions between particles.

OUTGOING ELECTRON

INCOMING ELECTRON

Feynman diagram

This Feynman diagram is used to represent the process of electron–electron scattering, with the repulsive electromagnetic force between them mediated, or conveyed, by a photon.

TIME

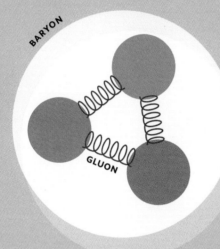

MESON (PAIR OF QUARKS)

A meson is made up of a quark–antiquark pair
(see pp.128–129), bound by the strong force.
The quarks have opposite color charge,
resulting in an overall colorless particle.

MESON

BARYON

GLUON

GLUON

BARYON (ODD NUMBER OF QUARKS)

A baryon is made up of three quarks. For
example, protons are consists of two up
quarks and one down quark bound by the
strong force. All three primary colors
(red, blue, and green) are represented,
making it "colorless" overall.

THREE-COLOR QUARK

Quantum chromodynamics (QCD) is the quantum field theory for the
strong interaction, which involves the exchange of gluons between
quarks (see p.122). It has many parallels to QED (see opposite), with
color instead of electric charge and gluons instead of photons.
However, the strong interaction has some unique characteristics,
resulting in behavior such as color confinement (which means that
color-charged particles cannot be found on their own) and an
extremely limited range of approximately 10^{-15} m.

QUAN
GRAV

T U M
I T Y

In the 20th century, two revolutionary theories emerged in physics: General Relativity and quantum mechanics. General Relativity describes physics on astronomical scales and models gravity as a geometric property of space-time, as it warps in the presence of mass and energy. Quantum mechanics describes physics on a scale in which gravity appears insignificant and inexplicable. Physicists hope to reconcile these by describing gravity according to the principles of quantum mechanics. The two most popular quantum gravity theories are string theory and loop quantum gravity (which does not treat gravity like other fundamental forces).

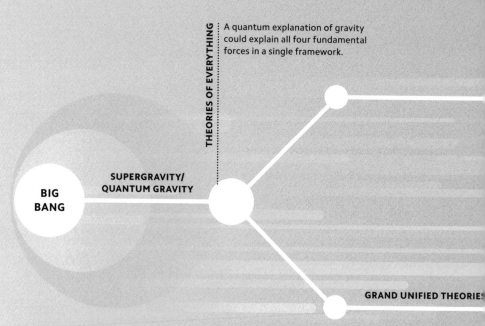

THEORIES OF EVERYTHING

A quantum explanation of gravity could explain all four fundamental forces in a single framework.

SUPERGRAVITY/ QUANTUM GRAVITY

BIG BANG

GRAND UNIFIED THEORIES

All in one
A theory of everything predicts that, at exceedingly high energies—such as just after the Big Bang—the four fundamental forces are united into one "superforce."

COMBINED FORCES

In the 20th century, physics coalesced into two frameworks: quantum mechanics and General Relativity. If these could be unified in a single theory, it would describe all phenomena in the universe: a "theory of everything." This is a Herculean task because according to General Relativity, gravity is not a force, but a property of space-time (see p.23). All efforts to model gravity as a quantum force (like other fundamental forces) have fallen short.

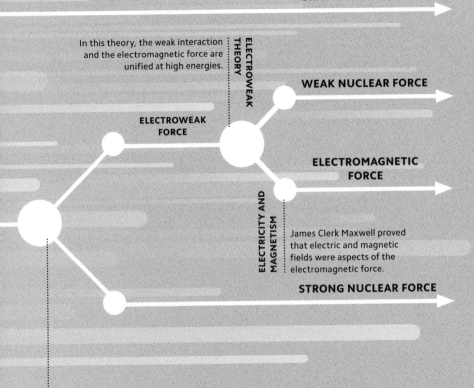

GRAVITATIONAL FORCE

In this theory, the weak interaction and the electromagnetic force are unified at high energies.

ELECTROWEAK THEORY

WEAK NUCLEAR FORCE

ELECTROWEAK FORCE

ELECTROMAGNETIC FORCE

ELECTRICITY AND MAGNETISM

James Clerk Maxwell proved that electric and magnetic fields were aspects of the electromagnetic force.

STRONG NUCLEAR FORCE

GRAND UNIFIED THEORIES

These seek to describe how all the forces except gravity can become one force at extremely high energies.

> "From time immemorial, man has desired to comprehend the complexity of nature in terms of as few elementary concepts as possible."
> Abdus Salam

QUANTUM FOAM

Under extreme conditions, such as in the center of a black hole or an instant after the Big Bang, the tried-and-tested laws of physics (General Relativity and quantum physics) break down. It is in this realm—the Planck scale—that physicists expect quantum gravity to emerge. The extreme quantities associated with this scale are measured with Planck units such as Planck length (~10^{-35} m). One unit of Planck length in comparison to the size of an atom is similar to the length of a football field relative to the entire visible universe.

NO SUCH THING AS EMPTY SPACE

On the Planck scale, space-time (see p.23) may not be smooth and empty but instead be full of fluctuations.

"In any field, find the strangest thing and then explore it."

John Wheeler

Space-time bubbles

Quantum gravity models predict that space-time is made up of tiny regions where dimensions froth in and out of existence, like bubbles in foam. This may be permitted by the uncertainty principle (see pp.42–43) over distances and intervals on the Planck scale.

Closed vibrating string
A closed string is a loop without any end point.
Closed loops are included in all string theories.

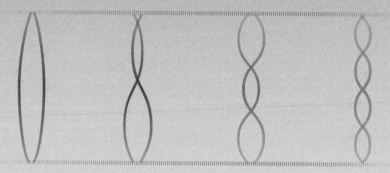

Open vibrating string
An open string has end points that connect
to 2D surfaces called d-branes. Not all string
theories incorporate the idea of open strings.

TINY VIBRATING STRINGS

String theory is a candidate TOE (see pp.138–39), which proposes
that all particles are vibrating one-dimensional strings. Different
vibrational states cause them to manifest as different particles,
including a force carrier for gravity (graviton). In string theory,
gravitons escape to higher dimensions, causing gravity to appear
very different from the other fundamental forces. There are several
different versions of string theory.

TYING IT TOGETHER

M-theory unifies five versions of string theory—which allow different types of strings—into a single theory. M-theory proposes fundamental building blocks called "branes," which are multi-dimensional versions of one-dimensional vibrating strings. The theory requires 11 dimensions (ten of space and one of time), and seven of them are "curled up" so small that they are invisible to us living in the other four dimensions. String theories, including M-theory, have been criticized for being possibly unfalsifiable.

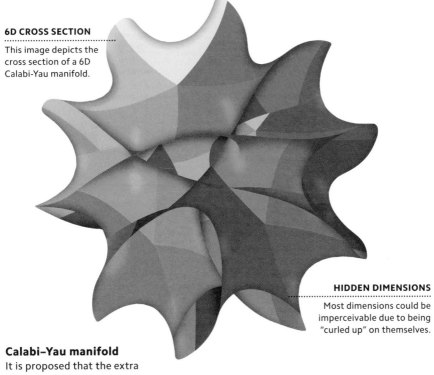

6D CROSS SECTION

This image depicts the cross section of a 6D Calabi-Yau manifold.

HIDDEN DIMENSIONS

Most dimensions could be imperceivable due to being "curled up" on themselves.

Calabi–Yau manifold

It is proposed that the extra dimensions of string theory are curled-up complex multidimensional space, such as this Calabi-Yau manifold.

THE FABRIC OF SPACE

Loop quantum gravity is a candidate TOE (see pp.138–39), which—rather than trying to unify the four fundamental forces— models gravity as a property of space-time (see p.23). The theory proposes that space-time is granular; it is built of Planck scale-sized "loops" of gravitational fields. Loops are woven into structures called spin networks, which represent states and interactions, and become a frothing "spin foam" when observed over time.

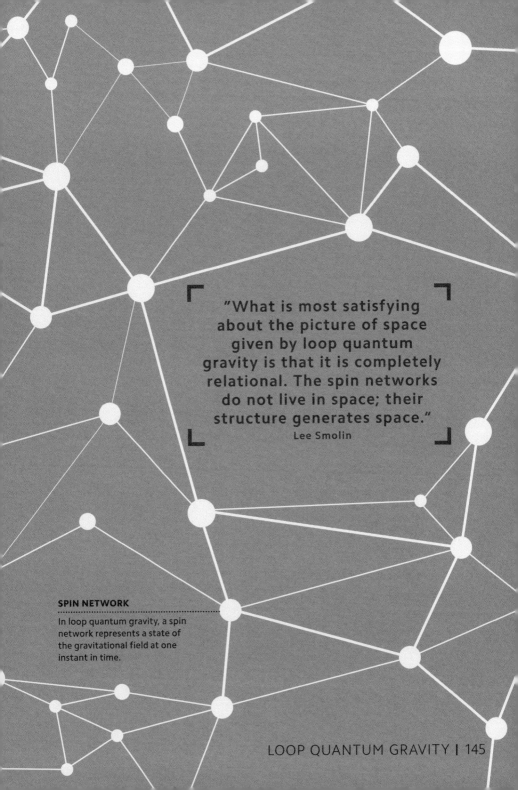

"What is most satisfying about the picture of space given by loop quantum gravity is that it is completely relational. The spin networks do not live in space; their structure generates space."
Lee Smolin

SPIN NETWORK

In loop quantum gravity, a spin network represents a state of the gravitational field at one instant in time.

QUAN
BIOL

TUM
OGY

At the most basic level, all the processes that make up what we call life come down to biochemistry—chemical interactions between a variety of complex molecules. Perhaps it should not be surprising, then, to learn that quantum effects have a part to play. Many pioneers in the field, including Erwin Schrödinger and Niels Bohr, predicted that quantum phenomena would play important roles in processes ranging from the harvesting of energy to genetic mutation, but it is only in the past few decades that our understanding of biochemistry has been able to reveal some of the details.

PHOTONS AND FOLIAGE

Photosynthesis is the process by which plants manufacture sugars and other chemicals using energy from sunlight. Its first step involves photons triggering chemical changes to molecules called chromophores. Energy produced by these changes is transferred to other molecules, where it can be put to use with remarkable efficiency, involving synchronized vibration between different energy states. Many biologists believe that photosynthesis has evolved to take advantage of the quantized nature of light energy, and some go further, suggesting that other quantum phenomena, such as superposition (see pp.38–39), may play a role.

SUNLIGHT

Inside the energy factory
A leaf is an electrochemical power plant that harnesses sunlight of specific wavelengths to trigger excitation in chromophores, and ultimately chemical changes in its pigment molecules.

PHOTONS

Incoming light carries energy corresponding to its color. Chromophores can be excited by high-energy violet and blue photons and low-energy red ones, but midrange green light is reflected back.

SHORTEST ROUTE

LONGEST ROUTE

Changes to pigments produce separated positive and negative ions, creating an "electrochemical potential" that can trigger other reactions.

REACTION CENTER

LIGHT-COLLECTING MOLECULES

Energy is transferred via neighboring chromophores to the reaction center with high efficiency. Each molecule in the chain is always ready to receive the energy in turn, so it is suggested that quantum phenomena aid the cells in finding the most efficient paths.

IT'S IN OUR DNA

The process of mutation plays a vital role in evolution by introducing random changes to DNA, the self-replicating biochemical molecule that carries instructions for making living things. Mutations involve individual chemical units called "bases" (the individual letters of the DNA code) abruptly changing from one form to another. Because bases seem inherently stable, some scientists think that quantum tunneling (see pp.70–71) is needed to leap energy barriers within their structure and allow them to change.

DNA consists of a chain of base pairs linked by sugar-phosphate "backbones" that twist into a spiral, or helix, shape.

DOUBLE HELIX

DNA STRAND

BASE PAIRS

Bases bond in specific pairs—adenine to thymine, and guanine to cytosine.

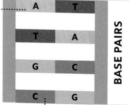

BASE PAIRS

REPLICATION

Quantum change

One form of mutation may start with a quantum event in which a proton from one base tunnels to its neighbor. This alters the length of the bond between them and triggers an error when the DNA is replicated.

Base pairing ensures that DNA can be replicated by "unzipping" and rebuilding the opposite strand.

SUBSTRATES

The chemicals involved in the reaction may initially be drawn to separate areas of the enzyme surface by weak attractive forces.

ENZYME

As the substrates bond to the enzyme, chemical changes allow them to overcome the energy barrier that prevents a reaction, forming a product that is then released.

ACTIVE SITE

Catalytic process

Catalysis involves changing the structure of two or more chemicals to allow a reaction between them. In some cases, the barrier to reaction appears to be overcome only through quantum tunneling.

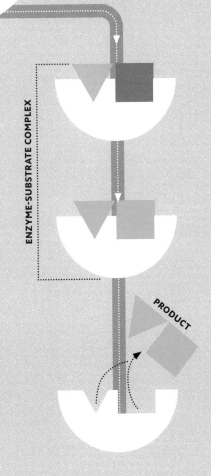

ENZYME-SUBSTRATE COMPLEX

PRODUCT

TUNNELING FOR A REACTION

Chemicals called enzymes are found throughout our bodies, aiding processes such as digestion (the breaking down of food into useful nutrients). They are thought to work by lowering the energy barrier between molecules called substrates so that they undergo reactions, but the precise way in which they do this remains poorly understood. One theory is that they do this by creating conditions in which electrons are able to bridge the gap between molecules using quantum tunneling (see pp.70–71).

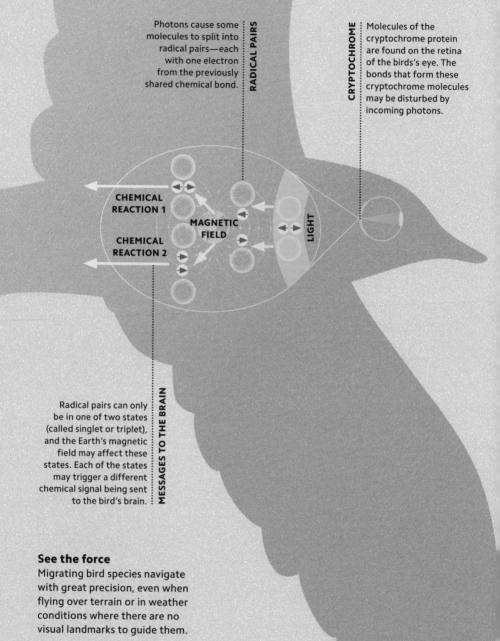

RADICAL PAIRS
Photons cause some molecules to split into radical pairs—each with one electron from the previously shared chemical bond.

CRYPTOCHROME
Molecules of the cryptochrome protein are found on the retina of the birds's eye. The bonds that form these cryptochrome molecules may be disturbed by incoming photons.

CHEMICAL REACTION 1

CHEMICAL REACTION 2

MAGNETIC FIELD

LIGHT

MESSAGES TO THE BRAIN
Radical pairs can only be in one of two states (called singlet or triplet), and the Earth's magnetic field may affect these states. Each of the states may trigger a different chemical signal being sent to the bird's brain.

See the force
Migrating bird species navigate with great precision, even when flying over terrain or in weather conditions where there are no visual landmarks to guide them.

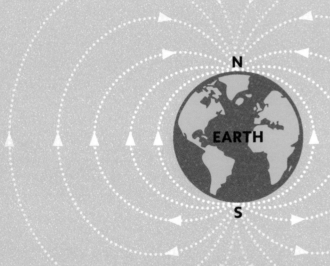

MAGNETIC FIELDS

N

EARTH

S

MAGNETIC PERCEPTION

Seasonal migrations see many different bird species fly vast distances between their winter and summer habitats. Experiments have demonstrated that birds rely on some sort of internal "compass" to navigate, and some scientists have proposed explanations that quantum effects are responsible. Proteins called cryptochromes found in the retina of the eye form pairs of molecules with correlated spins (see p.66) in blue light, and these spins can be oriented by magnetic fields, perhaps allowing birds to see Earth's magnetism.

OUR SENSE OF SMELL

The traditional understanding of our sense of smell, known as the "lock and key" model, involves a scent molecule (odorant) fitting into a receptor cell in the nose and triggering a sensory response. But is that the whole story? The unproven "vibrational" theory of olfaction uses quantum effects to offer a solution to some outstanding questions. It suggests that our odor reception involves a quantum tunneling effect, driven by the vibrations of scent molecules.

OLFACTORY BULB

The molecular structure of odorants is not static, but instead vibrates rapidly, emitting infrared energy in different "modes" akin to musical harmonics.

ODORANT MOLECULES

Quantum-tuned for smell?
According to one model, signals to our brain are triggered when electrons from an odorant molecule tunnel into our receptor proteins, boosting these complex molecules from one energy level to another.

QUANTUM CONSCIOUSNESS?

Conscious thought appears to be a uniquely human ability—but could our ability to reason, imagine, and assess problems be rooted in quantum physics? Several renowned physicists have argued that the unique aspects of our brains might arise from the harnessing of quantum phenomena such as entanglement (see pp.72–73) and superposition (see pp.38–39), but others doubt that quantum uncertainty could be sustained for long enough in our warm, wet bodies for any brain function to take advantage of it, due to decoherence (see p.75).

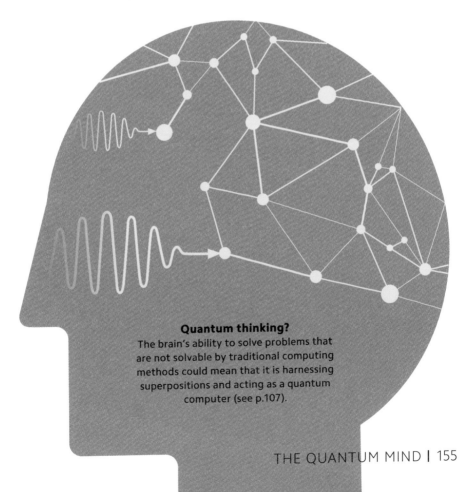

Quantum thinking?
The brain's ability to solve problems that are not solvable by traditional computing methods could mean that it is harnessing superpositions and acting as a quantum computer (see p.107).

INDEX

Page numbers in **bold** refer to main entries.

ACKNOWLEDGMENTS

DK would like to thank the following for their help with this book: Dominic Walliman for work on the contents list; Mik Gates, Dan Crisp, and Dominic Clifford for illustrations; Katie John for proofreading; Helen Peters for the index; Senior Jacket Designer Suhita Dharamjit; Senior DTP Designer Harish Aggarwal; Jackets Editorial Coordinator Priyanka Sharma; Managing Jackets Editor Saloni Singh.

All images © Dorling Kindersley Limited